HOW TO INV

HOW TO INVENT

M. W. THRING
Professor of Mechanical Engineering,
Queen Mary College

and

E. R. LAITHWAITE
Professor of Heavy Electrical Engineering,
Imperial College of Science and Technology

M

First published 1977 by
THE MACMILLAN PRESS LTD
London and Basingstoke
Associated companies in New York Dublin
Melbourne Johannesburg and Madras

ISBN 0 333 22026 9 (hard cover)
    0 333 17794 0 (paper cover)

Filmset in Ireland by Doyle Photosetting Ltd., Tullamore
Printed in Great Britain by Unwin Brothers Limited,
The Gresham Press, Old Woking, Surrey,

CONTENTS

PREFACE

Society needs good inventions more than ever before as the world's resources become scarce while one-third of the world's rapidly increasing population is undernourished. Inventing means applying a principle which is essentially different from those so far used on a particular problem and which is not derivable by a unique logical process.

The word 'invention' is essentially linked to the word 'new'. Patent literature is full of the words 'novel' and 'novelty', for newness is the essential ingredient to any valid patent. Even within the legal framework of Patent Law the idea of *newness* extends to a broader field than merely new pieces of hardware, for patents can be granted for new processes. Where inventiveness and the legal interpretation of it are different is that you cannot patent ideas (lest they never be fulfilled in practice) nor applications of a known idea.

In the broader definition of invention, however, it can be said to consist of a *new idea about anything* and therefore a comedian invents new jokes, a clown invents a new funny walk, a detective-fiction writer invents a new plot, and so on. But in this book we shall largely follow the legal definition and confine our objectives to new and useful pieces of hardware.

The primary objective of this book is to give people who have a spark of creativity in them the possibility of using this spark to invent practical useful objects. It is our experience that far more people have the potentiality of inventing than ever learn how to develop and use this power and, indeed, that much of our present educational system works against such development because it tends to make the student feel that he cannot achieve anything that has not been done much better before. We believe that this is a very harmful situation and we try to show the reader how to pinpoint some of the many small and large problems of extremely positive human value which are available (chapter 3) and how he or she can set about solving the ones they feel are really important.

One cannot teach creativeness, but the surprisingly large number of people who have it can learn how to direct it to the effective solution of the problems they have chosen.

We show how, in order to be able to invent, it is necessary to train the three 'brains': (i) the emotional brain (chapters 3 and 4) to give the necessary strength of purpose, determination and persistence, (ii) the intellectual brain (chapters 5 and 6) to ensure that one's inventions obey the laws of science and to be able to use analogical thinking, and (iii) the physical brain (chapter 7) to ensure that the inventions can be turned into real operating hardware. In chapter 10 we assemble some of the techniques we have found helpful in arriving at the inventive moment.

Teachers who are trying to encourage and develop inventiveness in their students may find chapters 2 and 8 helpful. People who want to find how to produce inventive solutions to problems that are around them are advised to look at chapters 4, 5 and 6 and to think about problems outlined in chapter 8. Chapter 4 gives our views on the actual technique of reaching the inventive moment. Readers who are interested in the idea and philosophy of invention may find chapters 1 and 2 interesting. We hope that the practising engineer and applied scientist may find chapters 7 and 9 of value in converting ideas into working reality.

We do not deal with economics in detail because we have concluded that premature entry of economic factors into the technological statement of the problem can be entirely inhibiting to the creative step of invention. Indeed, it can be demonstrated that most of the really big steps forward, such as tonnage steelmaking and the steam turbine, would have certainly been judged hopelessly uneconomic if a committee had been assessing them at any stage before the subsidiary inventions and prototype development had overcome the economic obstacles. They were so judged by all the experts in contemporary technology and this is why we encourage the would-be inventor to judge for himself or herself and not to be put off by the experts.

London, 1976   M.W.T.
                                                                                          E.R.L.

1 WHAT IS 'INVENTION' AND CAN IT BE TAUGHT?

M. W. Thring

YOU CAN BE AN INVENTOR

It is a common fallacy to believe that only one person in several thousand is capable of invention. On the contrary most people at some stages in their life perform genuinely inventive acts; what is rare is for these inventive acts to lead to a new device, process or product which is commercially viable. Whenever we solve a small problem which has no obvious logical solution, by finding an unexpected solution, we are inventing.

We can call 'invention' the conception of an idea which is later put into hardware to solve a practical human problem or satisfy a human need in a way which is not an obvious extension of known methods; there is always a quantum jump in ideas. It is not a greater number, size or the use of a stronger material, but it involves a different basic design principle. It is this quantum jump to a new principle which could not have been derived from the existing method by any logical process which constitutes the creative act of invention. The fact that someone else has invented a similar solution before does not stop it being an invention if the inventor was unaware of this and arrived at it by his own independent mental process.

Housewives commonly think of new and more labour-saving ways of arranging their kitchens, interesting new ways of planning their house-holds; gardens are full of inventive acts; for all creative artists invention is a key part of their work just as much as professional technique; the office worker can invent a better way of filing material for ready access; most home workshops are full of original inventions to make the tools and materials accessible and the particular type of work convenient within the limitations of the space available. Even in administration and human relations inventions can solve what appear to be insoluble problems. When we have judged children's invention competitions we have been immensely impressed by the scope and range of the original ideas invented, often in fields where we have ourselves worked and we say, 'Why didn't I think of that myself?', a criterion applicable to all the best inventions.

One cannot of course take people at random, or even people chosen from first class honours graduates or those with the highest I.Q., and teach them to be good inventors. The basic quality of being capable of really original ideas is of course part of a person's make-up and cannot be imparted if it is not there at all. But this quality is more common than one would guess if one judged by the proportion of people who achieve original results in our present society. One can draw a useful analogy with success in athletics of all kinds. The marathon runner must be a wiry man and the Olympic shot putter a massive man, but it does not follow that because a man is born with big bones and the ability to grow big muscles he will necessarily win the shot putt. He must be highly trained, physically balanced, exercise to develop particular muscles and acquire bulk. He must have coaches – and while *they* need not be big shot putters themselves, it helps a lot when it comes to the final stages if at least one of his handlers has had competitive experience. More than any of these, he must have a firm belief that he can win and an unshakable determination to do so.

We have come to the conclusion that the ability to do well in university science or engineering exams is totally uncorrelated with the ability to have original inventive ideas. We believe that people who get first class honours degrees in their finals are neither more likely nor less likely than university failures to produce such ideas. It is true that the examination-weak students are more likely to put forward ideas which are contrary to the known laws of Nature (for example, perpetual motion machines) whereas the well-drilled student will reject such ideas without mentioning them to anyone else, but the latter also tends to be inhibited from original thought by excessive addiction to analysis and excessive respect for the authorities in his subject.

A very good illustration of the fact that the ability to invent practical devices is latent in people who have not used it is provided by the work of physicists in Britain in World War II. Until the war most of them lived in the classical world of physics expressed in Rutherford's words 'It will take all the fun out of it if anybody finds a use for it.' However, when the war forced them to invent solutions to wartime problems they came up with radar, degaussing, infra-red detection and played a major part in the production of the atom bomb.

There is little doubt that the natural ability to have ideas which are genuinely original to the person concerned is quite widespread, indeed probably most people of normal intelligence have it. By 'genuinely original' is meant that the inventor has never seen or heard of this solution to this problem – often it has of course been previously invented by other people in the same or closely similar form but this was not known or if it had been, it was not the source of the idea in the inventor's mind. Why then do people so rarely make more than comparatively trivial localised use of this ability?

EDUCATION AND INVENTION

The answer to the foregoing question lies largely in three major defects of the education systems of all developed countries. These defects occur practically to the same degree in all countries with a formal universal education system, regardless of historical antecedents and political creed.

Education should be the complete development of a full man for a full life. A properly educated person would be one whose three faculties – called in poetic language, head, heart and hands – were trained and self-developed to their fullest possible capacity. In scientific language, these faculties are referred to as intellect, emotions (or feelings) and physical skills or techniques. Most education systems follow a scheme based upon ideas put forward by Plato; in these systems those people who can achieve above average intellectual achievement have a concentrated intellectual training and look down upon the manual craft skills for which their education finds little time. In Britain the onset of examinations drives out all free creative manual activities as the unfortunate child enters his or her teens.

The defect of the education system with regard to physical skills is therefore that it tends to despise them as only fit for non-intellectual beings, so that we can hear people who pride themselves on their scholarship saying, without any shame, 'I cannot knock a nail in.' We shall see in chapter 7 that the knowledge of reality which can only be obtained by skilled working with the hands is as essential to effective invention as the scientific experiment in the laboratory is to the development of science. We express this in the phrase 'Thinking with the hands'. It follows that if effective inventiveness is going to be experienced by everyone with the latent capacity for it, then everyone should learn at least one artistic or craft skill as well as they are able, as part of their normal education. Not only would this open up to them the possibility of making their ideas effective in the real world but it would also enrich their whole lives by enabling them to enjoy the pleasure of skilled creative work with the hands – certainly one of, if not the most fulfilling of all activities.

Even intellectual education suffers from one common defect, for students of the humanities (in spite of their claims to teach and encourage originality) as well as mathematics and the sciences. There is a tendency for teaching to consist of transmitting facts, dogma, opinions, ideas, theories and theorems to the student who then has to regurgitate them in half-digested form in examinations. It is rare for a really good teacher to teach the student to think for himself and work out his own ideas, opinions and conclusions or for science to be taught in such a way that the student is led to make his own discoveries, conclusions or hypotheses. Yet this is what the word *educare* means: 'to draw out' from the student. This has two serious harmful effects. First, the student rarely has the feeling that the acquisition of knowledge is an exciting process; the second is, from our present point of view, even more destructive. The student fails to acquire any belief that he himself is capable of original thought. As we shall see it is this self-confidence which is the first requirement in an inventor. This could only be put right by keeping an hour or two a week free from the examination cramming process and using it for open-ended problems for which every student is expected to come up with a different solution and for free ranging discussions and arguments in which students invent their own problems.

One can illustrate the discouragement faced by someone who has an original idea by some examples. Suppose I put forward a new idea that people could propel themselves several times as fast as they can walk and expend less energy by balancing on a two-wheeled device worked by pedals. I should be told 'We are not acrobats' if the bicycle was not familiar to us. Then I might suggest that we could go even faster without using our muscles at all, by using a machine in which thousands of explosions a minute did the work. Again I would be greeted by a chorus of horror at the dangers and impracticability of such a machine.

In chapter 8 we suggest ways in which originality can be encouraged and developed in the young inventor and in chapter 9 how an original idea can be turned into practical hardware.

The education of the emotions is by far the most difficult problem in education and one from which our present system has opted out completely. It is true that by studying great works of art or literature or the lives of great scientists or inventors or by working with a really good teacher a certain training of the emotional brain may rub off accidentally. But it is the emotions that provide our motivation – the driving force that we need to compel us to make efforts against our natural inertia and laziness. Indeed, in the Eastern analogy for man's three 'brains', the body is called the cart, the intellect the driver and the emotions the horse that pulls the cart. Educating the emotions would be equivalent to giving the driver reins so that he could control the horse's movement. Anything really worth doing, like inventing something of real value of humanity, is like pulling the cart out of a bog and up a steep hill – it requires all the effort the horse can make, the most determined control of its direction by the driver and the utmost self-confidence that such a task can be achieved.

The task of educating the emotions is naturally the most difficult part of a proper education of the whole man, and it is our failure to give young people an adequate ability to motivate their lives in a worthwhile direction that is one cause of the ever-increasing problems of the affluent society. Among these problems one can quote pollution, unemployment, unfulfilling work, the plight of the less developed countries, squandering of the earth's limited resources without regard for the needs of future generations and the arms race. All these lead to a steady deterioration of the quality of life of the individual so that a proper education in motivation must start with a clear explanation of the fallacy of confusing the standard of living, which is a cake of limited size in a crowded world, and quality of life, which has the characteristic that if one person enjoys more of it, then so does everyone around him or her. Such understanding of the basic difference between these two criteria of success in life can lead to the would-be inventor developing a determination to invent machines that increase the quality of life of the individual and do not add to the ever-increasing problems of the affluent society. In chapter 3 we shall try to identify the fields in which the inventor who sees this distinction clearly can have the greatest probability of making a real contribution to the quality of life of the citizen of an overcrowded world.

INVENTION AS A CREATIVE PROCESS

It is a basic hypothesis of this book that creativity in general and invention in particular come under a law which is neither causal nor casual. We assume in fact that human beings live at different moments of their lives under three different laws.

(1) *The law of accident* – randomness, casual events resulting from the interaction of quite different sequence lines of events with one's own sequence line. The extreme example is a brick falling on one's head as one walks along a street. Physics accepted this law in the first quarter of the century when quantum mechanics was developed to supplement the causal laws of classical physics. This law can be fed into a computer.

(2) *The law of causality* – the logical consequence of one's actions. If I eat too much food I get fat; if I drink too much alcohol I get drunk. This is the law on which computers operate, at least when there are no malfunctions. In the human mind it is the logical pursuit of a line of thought, so objective that all people who do it properly arrive at the same conclusion. All the laws of classical physics come under this heading.

(3) *The law of free will* – the process by which a human being makes a free choice or decision which could not have been predicted because it is neither a logical consequence of its antecedents nor an accidental random process. Careful introspection will convince anyone that at least in small things he does have a real freedom of choice, although he very rarely exercises it. Many words in the English language relate entirely to people acting under this third law, for example courage, self-discipline, self-control, decision, perseverance, effort, inner struggle. The only entry of this law into physics is the conception of Maxwell's Demon which could reverse the increase of entropy by opening and closing a little door to separate gas molecules into higher and lower temperature groups. This idea is very characteristic of the law of free will which does enable a creative person to create order out of chaos. It is not of course realisable because human consciousness, which is necessary for free will, cannot exist on a sufficiently small scale.

As soon as one accepts the possibility of a human being living occasionally under the law of free will, the whole human situation is changed, because one can make small decisions which gradually change oneself to the point where one can have the power to make greater decisions. Just as an artist prepares himself to paint the kind of creative pictures he wants to paint, so the would-be inventor can prepare himself to invent the kind of things he wants to invent; this book is concerned largely with developing the methods of such self-preparation.

The pure scientist can occasionally achieve one of the truly creative acts of

(1) producing a new hypothesis which leads to a deeper understanding of a body of scientific observation or (2) perceiving that an unexpected observation is not due to the cussedness of nature trying to spoil his experiment but to an important, hitherto undiscovered phenomenon. The true artist strives for creative acts in all his work but achieves it only occasionally. The engineer strives for the truly inventive idea which creates a new order in the solution of a practical human need.

WHAT IS INVENTION?

Arthur Koestler in various books, but especially in *The Act of Creation*,[1] has analysed truly creative ideas by the expression 'bisociation', which he defines as the solution of a problem in one matrix of associated ideas by bringing in an idea from an entirely unconnected matrix. Inventing a new process or product is certainly a truly creative act just as is the invention of a new hypothesis that makes sense of a group of scientific observations, and Koestler's analysis certainly applies to the act of invention. The example of Archimedes jumping out of his bath and shouting '*Eureka*' because he had thought of a way of measuring the volume of the king's crown, and so deciding whether it had the density of copper or of gold, is a well-known story of an invention to solve a practical problem by applying to it a hitherto unconnected scientific observation. Many, if not all, true inventions correspond exactly to bisociation in the sense that they correspond to solving an apparently insoluble practical problem by bringing in an idea which is well known in some quite other connection, but which could not be reached by a direct logical process or a systematic search of a predetermined field of knowledge such as a computer can be instructed to perform.

A closely similar idea has been expressed by Edward de Bono in the concept of 'lateral thinking' which implies a thought process which proceeds sideways from the normal branched chain of logical possibility. This he illustrates very nicely with the action of a girl forced to choose a pebble from a bag which is supposed to contain 3 black and 3 white pebbles, when she finds out that they are all in fact black. She picks one out and drops it, to disappear in the gravel; she then claims it was white and the villain is forced to agree or his deception will be exposed.[2]

In our view both these descriptions apply only to the intellectual aspect of invention, whereas a true invention involves the complete working together of the three 'brains' of the inventor. A real invention in the sense the word is used in this book is the conception that can lead to a device (for example, mechanical, electrical, electronic) which can be constructed and worked to serve a human need of some kind in a way which is clearly better than before. Two other essential parts of an invention are therefore (1) the strong feeling of desire to produce such a better solution to a human need and (2) the understanding of the way things work in space and time through the hands and eyes without which no realisable idea can be born (see figure 1.1).

That the emotional 'brain' is essential to the achievement of a successful

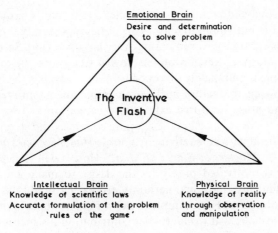

Figure 1.1   The act of invention; the three brains work

invention is clear to everyone who has invented anything, however slight. An inventor must start with the absolute certainty of belief that he can succeed, in spite of the fact that everyone else will tell him that he is bound to fail when so many better men have tried and failed. He must also have the absolute emotional certainty that the solution of the problem is important enough to enable him to produce the necessary emotional power. He must suggest not merely one really creative solution, but often a whole series of original ideas, since the majority will have to be rejected because they prove either to be completely unworkable or to produce undesirable side effects which nullify their value. Again an idea may seem to be unworkable but a further emotional effort may result in an associated invention which overcomes the objection to the first. The would-be inventor must necessarily pass through black periods when even to himself the problem seems insoluble, or when a cherished idea has on further consideration proved to be worthless. Success is achieved only if one has the emotional strength necessary to continue the struggle to find a solution even when it seems hopeless. Not even the greatest inventors achieved their successes without such difficulties, and knowledge of this fact can be a great comfort to the young man determined to find an inventive solution to a problem.

This essential emotional content of invention means that the preparatory step is not only to work out what is the exact problem you want to solve but also to make up your mind that it is really worth solving. It is no longer sufficient to feel that one can make a personal fortune by an original invention for two reasons. First, the kind of invention that can lead to a personal fortune is, to say the least, thousands of times more difficult to achieve than it was in the nineteenth century. Then, the rapid expansion of the tools, machines and materials of the Industrial Revolution was constantly opening up new possibilities and the first inventor to exploit one of them had a fair chance of personal affluence. Now, even the rapidly developing subjects like electronics, solid-state materials, computers, and robots, require teamwork which gives the individual inventor very little opportunity to succeed in his own business without government or other large backing. Second, the flood tide of the Industrial Revolution has flowed to a point where that which is most economic is completely divorced from that which is the best servant of humanity, owing to factors such as unemployment, uninteresting work, pollution and noise, resource exhaustion, escalation of war weapons, the gap between the developed and developing countries and so on. This divorce of interest is dealt with at length in my books, *Man, Machines and Tomorrow*[3] and *Machines: Masters or Slaves of Man?*[4] Here it will suffice to say that if an inventor is motivated purely by the desire to make a fortune he will necessarily find his emotional certainty eroded by doubts as to the social value of his objective. For example, if he works on robots in order to reduce production costs he will wonder what contribution his success will make to unemployment. For this reason chapter 3 of this book constitutes an attempt to work out what inventions are most needed from the point of view of the whole of humanity now and in the future, and of the limited environment in which humanity lives – Spaceship Earth.

A reasonable analogy for the emotional aspect of invention is to consider the inventor as a man riding a spirited horse in a field surrounded by high hedges. He is determined to find his way out and he firmly brings the horse at a gallop to point after point of the hedge, and time after time the hedge is too high and the horse refuses or fails in an attempted leap, but if the rider's decision and determination are sufficient he can eventually find a low point in the hedge and get clear to gallop freely forward in a new field.

The importance of the physical aspect of invention is shown by a study of the lives of great inventors, some of which are discussed in chapter 2. All of them in their youth 'messed about' in workshops or home-made laboratories and did all kinds of self-planned experiments and made gadgets to solve problems of their daily life – Parsons' steam-operated pram was a rather extreme example! In this way they acquired the feeling for structural strength and rigidity. Buckminster Fuller's skill in inventing complex, rigid three-dimensional shapes (geodesic domes, 'tensegrity' masts and tetrahedral rigid floors) is an excellent example of the use of the ability to visualise inventively in three dimensions. This feeling for the way things work only enables the inventor to visualise devices which can work within the field of knowledge of which he has the key through his experiences and experiments. When I was a young student studying physics in the Cavendish Laboratory and heard Rutherford lecture I was most impressed by his ability to construct a mental model of the atomic structure based on his experience. The great contribution of the nuclear model of the atom came from this physical experience whereas Einstein's contribution to general relativity was based on the application of a mathematical technique to explain physical facts. This may perhaps be why he failed in his subsequent fifty-year attempt to produce a theory which would 'explain' electromagnetism as well as he had 'explained' gravity. This whole field of 'thinking with the hands' will be considered in chapter 7 with some methods of teaching it as a skill at school and university.

Figure 1.2 illustrates the role of invention in human knowledge. The vertical axis represents the fundamentality of knowledge, with the more fundamental understanding at the bottom and the more practical *ad hoc* and superficial at the top. The horizontal axis represents the extent to which the knowledge is orientated towards the satisfaction of human curiosity about the way the Universe and its components work (on the left-hand side) or towards the satisfaction of the physical needs of human beings – including the need for beautiful surroundings – on the right-hand side. It is often said that the classical procedure of pure science automatically throws up benefits to human life, but this is quite incorrect. The classical process of pure science is the path down the left-hand side of the diagram: first a group of observations of natural phenomena are made and a natural history is constructed from them. Then a hypothesis is put forward which can 'make sense' or put a complex mass of facts into a simple theoretical framework – for example Mendel's Law of Inheritance or Einstein's Special Relativity.

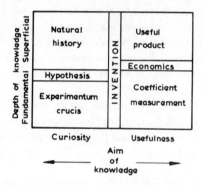

Figure 1.2   Plan of human knowledge

Then experiments are carried out to test a deduction from the theory, and the *experimentum crucis* either verifies or disproves the theory, at least for a while. It is only if an inventor sees that the scientific theory shows that a new way of satisfying a human need can be obtained by an act of invention that the science can bear practical fruit. Mendel's laws have been applied to produce strains of wheat, fruit and vegetables with valuable properties, but it required the intervention of an inventor who knew that these properties would be valuable. Special Relativity contained the idea that mass could be converted to energy but a practical realisation could not be envisaged until, more than thirty years later, the fission of the uranium nucleus had been achieved. Then it was possible to envisage a nuclear power station. I remember telephoning my former Cambridge Director of Studies during World War II to ask him if it would not be possible to obtain power from the nucleus because the Director of BCURA who employed me was preparing a paper on future sources of power. I received a very non-committal reply because he was himself working on the secret Tube-Alloys project (the code name for the atom bomb development work) at the time

The various ways in which an invention can take knowledge from the left-hand side of the diagram to the right-hand side will be discussed in chapter 7, but here it is necessary to strike a blow against the favourite snobbishness of the pure scientist. This is the idea that pure science is fundamental and that engineering, applied science and technology are superficial subjects concerned only with simple practical *ad hoc* problems. This idea is equivalent to ignoring the top left-hand and bottom right-hand squares of the diagram so that the pure scientist confines himself to the bottom left-hand square while he expects the engineer to confine himself to the top right-hand square. The inventor is not expected to understand fundamental theory at all, but only to invent practical devices based on superficial natural history-type knowledge. It is true that many successful inventors have done this but there are now many examples of cases where the inventor has started with the deepest scientific understanding and carried this through the fundamental engineering studies represented by the bottom right-hand square before reaching a practical device. The thermionic valve was based on an understanding of electron cloud behaviour in vacua, the Langmuir process of recombination welding on dissociation theory, the nuclear power station on measurements of the nuclear cross section for various process, and all transistor and other solid state electronic devices were based entirely on measurements of the required properties.

The bottom right-hand square represents the measurement of the physical coefficients and chemical behaviour of the materials involved in the proposed process. It is only by knowing these that the engineer can turn the inventive idea into a machine that can not only work but work continuously, economically and efficiently in space–time.

The diagram shows that the act of invention is the act whereby knowledge previously only of interest becomes capable of satisfying a human need.

THE QUALITIES REQUIRED IN AN INVENTOR

Obviously, as has been said above, a person with no spark of originality can never be an inventor. However at least 50 per cent of people of normal intelligence do have sparks of originality and all the other qualities required in an inventor can be developed in such people by suitable teaching and self-discipline. It is in fact far more important in becoming a successful inventor to have the qualities for overcoming obstacles, whether they be one's own ignorance, other people's obstructiveness or the 'cussedness' of inanimate objects, than it is to be constantly thinking of new and original ideas. The successful inventor is the man who sees clearly what is a truly valuable idea and then carries it past all the difficulties, often by means of subsidiary inventions, until it is fully realised in commercial hardware. This is what Edison meant when he said invention was 10 per cent inspiration and 90 per cent perspiration. Often the basic idea of the solution to a human problem is known to many people who are working on it simultaneously. Parsons was only one of those working on the steam turbine and his success was due to his being able to have his own machine shop as much as to his hunches about the best shape of the blades. The theory of steam flow in turbines (the fundamental engineering square in figure 1.2) came much later. Even more instructive is the fact that when in 1912 Lord Fisher asked Parsons to make a combustion gas turbine, Parsons said he did not think it was theoretically possible. He probably meant that he did not believe one could design a compressor efficient enough not to absorb all the power of the turbine. However it shows that the idea of the gas turbine was mooted nearly thirty years before Whittle brought it to actualisation. I remember visiting Firth Brown research laboratories in 1938 to ask about special heat resisting steels for my work on combustion and saying to them 'If you have such marvellous steels you could make a gas turbine' to which they replied 'We are'.

Thus the first quality required in a successful inventor is unbounded self-confidence amounting to brashness – he is quite certain as an act of faith that he can succeed and quite determined that he will. Such self-confidence exists in all young children who have been brought up by parents in whom the child's confidence of receiving love is fully justified. The trouble is that so much of education is designed to destroy this self-confidence and teach the child that knowledge is something which already exists and in which he has no novel part to play. Only by introducing open-ended problems of gradually increasing difficulty, so that the student learns that he can produce his own worthwhile solutions, can this self-confidence be restored. Chapter 8 is largely concerned with methods for restoring this creative self-confidence.

The second quality is also an emotional one; it is the perseverance and persistence to see the invention through every kind of obstacle and opposition and not to be put off by repeated failure. Some people have this stubbornness more than others, but every emotionally mature person is capable of it if he is sufficiently motivated by the feeling that success is really of deep value, not only to himself but to humanity. It is probably easier for a person who is naturally an introvert to pursue a goal that exists only in the mind's eye without excessive discouragement and despair at failure, but again there is little doubt that even a highly extroverted person can find the inner strength from a clearly worked out objective.

Third, the inventor needs to have developed in himself the intellectual and physical tools necessary to have a workable idea, the ability to think with the use of well-chosen analogues and models which retain the essentials of the problem but omit all the non-essential details, and the ability to 'think with the hands' so that complex moving systems can be conceived which can work. He does not necessarily have to possess great mathematical skill in setting down the differential equations and solving them because this can only be done after the new idea has been crystallised. Edison said 'I can always hire people to do the mathematics' – and did.

Fourth, the inventor has to have a good measure of self-knowledge; he has to know when to bully his mind into a determined attack on the problem and when it is better to let it lie fallow; he has to know what time of day and under what conditions his mind works most creatively, for example, in the middle of the night, after a few drinks or when teaching others about the problem; most important of all he has to learn the secret of switching off the critical faculty so that new ideas can form without being strangled by objections before they are fully in the mind – to get rid of the strait-jacket of inhibitions.

Finally, he must have what can be called the inventor's eye. This means he must look at every human device as if he were a man from Mars and say 'Why do they do it that way? What do they really want to achieve? Do they want to make so much noise and pollution? Could not the objective be obtained in some quite other way? Why is a cooling tower convergent – divergent; why do they have to evaporate water in cooling towers; why do some power station chimneys have four outlets; why do ships' propellers have three, four or five blades; why don't ships have four propellers; why did the propeller oust the paddle wheel; must trains have wheels; why is a car's exhaust underneath; why does a lavatory cistern use so much water; why do strip lights make so much mains-hum; must a vacuum cleaner be so noisy; do you have to put lead in petrol?' Any one of these and a thousand other questions can lead to an invention. It is in exactly the same way that the scientist looks at every phenomenon and says 'How can I make sense of that? How can I fit it in with the rest of my world in the simplest possible way?'

The inventor who has already decided that something could be done better, then looks at everything with this problem in the back of his mind and some small observation can give the clue to a solution. The classical example of this was James Watt, who wanted to find a better source of power than the windmill or the muscle, when he noticed the power of steam to lift the lid of a kettle.

REFERENCES

1. Koestler, A., *The Act of Creation*, Pan Books (1970)
2. De Bono, E., *The Use of Lateral Thinking*, Penguin (1971)
3. Thring, M. W., *Man, Machines and Tomorrow*, Routledge (1973)
4. Thring, M. W., *Machines: Masters or Slaves of Man?*, Peter Peregrinus (1973)

2. SOME HISTORIC INVENTIONS AND INVENTORS

M. W. Thring

THE BEGINNINGS OF TECHNOLOGY

The very first stage in technology was the use by man of hand held weapons and tools to increase his own physical abilities. The simple stone and the wooden club were followed by flint chipping, wood and bone shaping, bone needles with eyes and the invention that gave the greatest extension to man's strength, the wooden handle strongly fixed to a stone axe, spear or hammer head. Without this he could not cut down trees, dig standing up or hunt animals as large as himself.

Another invention of very great technical brilliance was the bow and arrow. The use of a light but strong string to harness the stored energy of a bent wooden spring to give a light arrow a high acceleration clearly involved creative thinking of a very high order and could hardly have occurred as a result of an accidental observation.

The use of controlled fiercely burning wood and charcoal fires later blown with skin bellows enabled man to smelt copper, bronze alloys and eventually iron which led to mediaeval armour, the sword (those with exceptional properties of hardness and toughness were sometimes regarded as magical) and the hand tools which achieved their summit in the felling axe, the adze and the craftsman carpenter's tools, many of which are still in skilled use. The combination of a light wooden handle with an iron head was a great invention brought from the Stone Age. The invention of the spade which enables a man to use the strength of his leg to force it into heavy soil made it possible to work the soil to the depth necessary to grow deep rooted vegetables. The earlier chopping hoe was only able to break the surface and was quite useless on the clay soils which cover large parts of Britain.

Probably the most significant invention of all, from the point of view of the ultimate development of technology, was that of the wheel. The Aztecs and the Incas managed to reach a high level of stone building without the wheel but only by a vast use of slave manpower. In Western civilisation the wheel existed by the time of the chariots of the Pharoah. The oldest known illustration is in a bas-relief from the city of Ur in Mesopotamia (third millenium B.C.). It was certainly invented in two stages; first the round roller cut from trees was used to drag heavy stones with much less friction than on simple sleds. Then some unknown genius invented the axle and bearing which has the double advantage of reducing the friction force leverage in the ratio of the axle to wheel diameters, and enabling one to choose both the bearing surfaces and smooth them to reduce friction. Mutton fat as a lubricant was probably used at a very early date to reduce friction still further. The spoked wheel appeared in Eastern Persia and in Thebes, Egypt, between 2000 and 1500 B.C. (*A Pictorial History of Inventions*, U. Eco and G. B. Zorgoli, published in English by Weidenfeld and Nicolson in 1962).

The other great technological developments prior to the first industrial revolution relate to the extension of the muscle power of the human, first by the use of draft animals and later the use of wind and water power. Many inventions were required before full use could be made of the great strength of the horse and bullock. First the saddle, and much later the stirrup, enabled a man to ride and swing two-handed weapons from horseback; the horseshoe protected the hoof on hard ground; and the horse collar enabled the full pulling power of the horse to be applied to the capstan, the cart and to the plough – itself a major invention which increased tenfold the area of land a man could cultivate.

The Greeks invented geometry and the world of ideas because they believed that free men should concentrate their intellectual activities on the discovery of the harmonious laws of the Universe. Thus they left mechanical activity and craftsmanship to the slaves and foreigners. We still suffer badly from what I call 'Plato's heresy', the contempt for the mechanic by the humanist philosopher and mathematician. Archimedes (287–212 B.C.) was a Greek mathematician who lived in Sicily and was forced to be an inventor by circumstances. He left no written record of his inventions and chose for his tomb a sphere inscribed in a cylinder to express his geometrical work. He is reported to have invented siege weapons, a burning mirror made of many small plane mirrors, the method of distinguishing gold from copper by measuring the volume by displaced water to give the density, the lever, the Archimedian screw and a planetary motion sphere.

Heron of Alexandria (150–100 B.C.) was also a Greek but he did write books on mechanical subjects, such as (i) the *Pneumatica* which describes fountains, siphons, a fire-engine and the aeoliopyle (a reaction steam turbine which spun when a fire was lit under it), (ii) a book on engines of war, (iii) *Mechanics*: which describes many machines such as a distance indicator for a cart which counted the wheel revolutions with multiple worm reduction gears, (iv) a book on water clocks, (v) a work on land surveying which uses an instrument, the dioptra, the forerunner of the theodolite.

Probably the next inventor whose inventions are well known was Leonardo da Vinci (1452–1519). He had an inexhaustible intellectual energy and curiosity and was a Master of several arts (painting, sculpture, music and architecture) as well as an inventor and scientist. He worked on architectural plans involving improved sanitary principles and plans for improved irrigation in the Lombardy plain, the canalisation and control of the waters of the Arno and the prevention of landslip. His scientific studies were closely related to his desire to invent solutions to practical problems or, in the case of his anatomical studies, to his painting. He does not seem to have suffered from Plato's heresy at all, probably because his first training was an apprenticeship to a man skilled in the crafts of the goldsmith, sculptor and painter. In the museum in Milan devoted to his life there is a whole gallery of models of the machines shown in his notebooks – machines for all kinds of manufacturing processes, for war purposes (a war tank propelled by hand-operated cranks but with the front and back wheels going in opposite directions; he also planned to use steam to drive projectiles from a gun) and human-operated helicopters and other flying machines with flapping wings for which he made an exhaustive study of bird flight. There is no evidence that any of his machines were developed to the point of actual use and some critics have said that certain sketches were copies of earlier inventions. Nevertheless his notebooks and sketches provide an immensely stimulating record of the working of an inventive genius, many of whose inventions were re-invented much later when the materials, the prime movers and the theoretical knowledge were available to make them feasible.

POWER FROM WIND AND WATER

The first waterwheels were simply a form of pump powered by human or animal muscles: rotation of the wheel lifted the water in a series of pots around the rim filled in the stream at the bottom to discharge at the top. Philo of Byzantium, a Greek, who lived in the latter half of the second century B.C. wrote a textbook on mechanics in which a wheel is described which had paddles of sufficient area to enable the flow of water to rotate the wheel and lift a small fraction of it automatically to the top. This beginning led to the undershot waterwheel and much later the overshot was invented for a smaller flow of water, but larger head. It is obviously in principle the same as the original wheel with pots but run as a power source instead of as a pump. These waterwheels provided power for corn-grinding mills (mill ponds are still a common feature of the British countryside); later one of the main reasons for the steel industry in Sheffield was the streams of water with considerable head coming from the Pennines.

The waterwheel has developed into the water turbine, which provides a useful proportion of the electricity in the developed countries, but requires both a great head of water and thousands of tons of water (e.g. 1 MWh requires 370 000 ton-metres of water) such as can only be provided when a lot of rainfall occurs in mountains above a reservoir, which is itself several hundreds of metres above the turbine discharge.

The first use of wind power, the sail, was much earlier. Nothing is known of the first inventor of the sail but the art of sailing developed over many centuries into a highly scientific body of knowledge which reached its peak in the nineteenth century. Sailing remains one of the most fascinating hobbies and it is possible that as oil fuel runs short in the next century we shall return to large sailing ships for cargo carrying and even passenger cruising using modern developments in materials – hulls and spars of carbon fibre and nylon sails, perhaps in the form of giant catamarans with partially flexible interconnection.

The windmill dates back to the early Middle Ages and depends on the ingenious invention that blades set at an angle to form a wheel with the axis pointing at the wind receive a steady thrust throughout the rotation. This is the forerunner of the propeller as it is based on the same principle.

The windmill also led to the first fully automatic control; a small windwheel was set at the back of a revolvable tower carrying the main wheel, with its axis at right angles so that whenever the wind shifted and had a component not along the axis of the mainwheel this small wheel revolved and drove the whole tower around until the main wheel was pointing in the new wind direction.

A number of clever inventions were made to prevent the blades of these windmills being torn off in high gales. For example each main sweep consisted of a large number of small blades which could be pivoted about an axis perpendicular to the sweep, like a Venetian blind, so that the sweep could be opened up to let the wind go through; a bar along the side of each could be operated by a crank on the central rotor axis. Many people have worked on the idea of a vertical axis windmill which works equally well whatever the direction of the wind; the same principle as the cup anemometer. The Savonius Rotor made of a 40-gallon oil drum split into two halves by two cuts parallel to the axis and set on a vertical shaft to make an S-shape is a practical modern development. I helped my father build one with automatically feathering blades which gave more area to the wind on one side than on the other. All these suffer from the defect that it is very difficult to support them strongly enough to withstand the occasional high winds, although my father's flapping one could have the flaps released to lie along the wind on both sides instead of only on the feathering side. The windmill, like the sailing ship, will see a great rebirth in the next few decades especially where it can be used for an energy storage system like pumping irrigation water to a high level pond on a farm or operating a heat pump to heat water.

THE BEGINNINGS OF THE FIRST INDUSTRIAL REVOLUTION

We shall pick out a few inventions from the period that started with the first industrial steam engines (the beam engines) and the use of coal gas for lighting, included the development of accurate machine tools, tonnage steel production, the paddle wheel and the screw propeller, the steam train, and concluded with the work of Parsons and Edison. This period very roughly coincided with the nineteenth century in Britain. This was the period when the individual inventor could obtain enough finance to build working prototypes and get them into industry if he was lucky. The objections to the new idea raised by the Government, the Admiralty, established businesses and workmen were just as obstructive as they are today, but the cost of financing a new development was within the means of a single business man (for example Boulton, who was the business partner of Watt) or of the inventor himself if he had had some previous smaller successful inventions (as Bessemer and Siemens). Edison and Parsons came at the end of this period but were able to raise enough money to do the development work which led to the establishment of large companies which are both still flourishing.

The original beam engine associated with the names of Papin and Newcomen in the period 1698–1908 used the same cylinder as the power cylinder and condenser. Newcomen introduced the use of a jet of water to condense the steam and an automatic device to work the valves. It was a tremendous advance on the windmill for pumping water but used an amount of coal corresponding to a thermal efficiency of just over 1 per cent.

Watt was a mathematical instrument maker to Dr Black, the Professor of Chemistry at Glasgow University, and repaired a model of a Newcomen engine in 1763. In 1765 he conceived the idea of a separate condenser so that the cylinder did not have to cool to the condensation temperature every stroke. Watt also introduced the ideas of cutting off the steam supply to the cylinder after a small part of the expansion stroke with a slide valve, so that the steam could do work as it expanded to a lower pressure, and the use of a double acting cylinder. He added a flywheel and a steam feed governor. Watt obtained capital to develop his machine first from Carron Ironworks, and then from Boulton, a Birmingham industrialist. By 1776 he had a machine in industrial use and he raised the thermal efficiency to about 8 per cent by 1800. His master patent on the separate condenser ran until 1800 and by then he had installed 100 pumping and 200 rotary power engines. He had to use a sun and planet gear to convert the beam oscillation into rotary motion because someone else had patented the crank.

The next step was taken by Trevithick who had an engine running by 1803 with 3 atm pressure steam. Watt called him a 'murderer' for using pressures above one atmosphere and he did have a boiler which blew up and problems with valves. Until the high pressure engine was combined with a condenser it had a lower efficiency than Watt's and indeed railway locomotives did not have condensers at any stage, but Trevithick's engine made the locomotive possible because the cylinder could be so much smaller for a given power. By 1835 the use of a condenser gave an efficiency of about 12 per cent while by 1917 the multiple expansion cylinder engine had increased this to 16 per cent.

About 1880 Otto and Diesel produced workable internal combustion engines which quite soon had a high thermal efficiency because the upper working temperature and compression ratio were very high in spite of returning to a temperature cycle in the cylinder itself.

William Murdoch was the effective inventor of gas lighting. He worked for Boulton and Watt and invented the sun and planet alternative to the crank in 1782. He superintended the erection and running of the steam pumping engines in the Cornish mines and built a model beam operated steam locomotive. It was known that one third of Newcastle coal was volatile and the vapours were inflammable. Murdoch heated coal in a kettle and burnt the vapours on a perforated thimble. In 1792 he lit his own house by gas; he used coal gas from a bladder for a lantern. In 1802 he gave the first public exhibition by lighting the Boulton and Watt works at Soho, although James Watt Jr had declined to patent the gas lighting system. Boulton and Watt accepted Murdoch into partnership and from 1805 onwards a number of mills were fitted with gas lighting and gas making retorts. Sir Humphrey Davy ridiculed the idea of street lighting by gas, asking if it was intended to take the Dome of St Paul's for a gasometer? In 1808 Murdoch was awarded a Royal Society medal for a paper 'on the application of gas from coal to economical purposes'. In 1814 the London and Westminster Gas Company lit Westminster Bridge by gas and during the next few years Glasgow, Liverpool and Dublin were lit by gas. Murdoch had no patent and made no money from this invention but managed Boulton and Watt's works and invented many new improvements for their steam engines and many other practical devices such as a cast iron camet, a vacuum method of transmitting packages through a tube, a compressed air motor and a pair of condensing engines with ninety cranks to drive a steamer.

Francis Pettit Smith ('Screw' Smith) was one of the many who invented the screw propeller for ships.

The account of the development of the screw propeller by Samuel Smiles (*Invention and Industry*, London, 1884) contains all the elements of a major invention:

(1)   An entirely different basic principle offering a big improvement.
(2)   Many people conceive the new system in principle but most of them abandon it at an early stage.
(3)   Severe and continued opposition by the official establishment only overcome by immense persistence by the inventor. It also contains an unusual element of lucky accident.

Steam driven paddle wheels were successfully used to propel ships in 1788 and 1801; a regular steamboat was on the Thames by 1815. James Watt sketched a 'special oar' in a letter in 1770. Joseph Bramah in 1785 patented ship propulsion by a rotary steam engine using either a paddle-wheel or a 'screw propeller' – similar to the fly (air screw) of a smoke jack. Screws were patented in the period 1790–1815 in Vienna, England and America. In 1834 F. P. Smith, a grazing farmer on Romney Marsh, aged 26, built a model boat propelled by a wooden screw driven by a spring, and concluded after trials on a reservoir that it was much superior to paddles. Two years later he patented vessel propulsion by a screw revolving beneath the water, and he was able to obtain financial support from a banker to build a 10-ton, 6 hp steam vessel which was fitted with a screw of two whole turns. He must have regarded the water as being like a soft solid such as clay through which the screw cut a thread. However the propeller struck an obstacle in the water and about one half of its length was broken off whereupon it shot ahead and attained a much greater speed than before. A new screw of a single turn was therefore fitted, but it is clear that a correct theoretical model of the action of a screw in thrusting a jet backwards was still far from the inventor's mind. Indeed Robert Wilson's 1827 model with 'revolving skulls' was much closer to the modern idea of a bladed propeller. However it was F. P. Smith who persisted in the development of the idea for a whole lifetime and who therefore must be regarded as the principal inventor. In 1837 he took his 10-ton vessel to the open sea, withstood severe weather and attained a speed of 7 mph.

Another inventor, Ericsson, the Swede, whose steam locomotive Novelty was runner up to Stephenson's Rocket in the 1829 competition, also patented a 'spiral propeller' in 1836 and built a 40 ft boat with two 5 ft propellers which reached 10 mph. This he demonstrated by taking the Lords of the Admiralty from Somerset House to Limehouse and back again, but their lordships expressed themselves 'very much disappointed with the results of the experiment'. This turned out to be because the Surveyor to the Navy had expressed the opinion that 'the power being applied to the stern, it would be absolutely impossible to make the vessel steer'. The U.S. Navy had two iron ships built in England with the Ericsson propeller and he went over to the United States; later he built the ironclad *Monitor* used in the Civil War for which he was not paid by the U.S. Government.

The British Admiralty tried 'Screw' Smith's ship in 1838 and it gave great satisfaction, but they insisted on a 200 ton vessel being built and tried. He was able to obtain finance to set up the 'Ship Propeller Company' which launched the 237 ton *Archimedes* in October 1838. This finally convinced the Lords of the Admiralty that a rear propelled vessel could steer. Even then the vessel had to be operated for many years before it was accepted by the Admiralty because Her Majesty's Principal Designer was opposed to iron ships, steam power and all other new projects. Several large passenger ships, notably Brunel's *Great Britain*, were using screws while the Admiralty was still conducting cautious experiments. 'Screw' Smith's patent expired in 1856 by which time 327 Naval vessels and far more merchant ships had screws. He made no money directly from his work but later received a subscription, a civil list pension and a knighthood.

The first molten steel was produced in 1750 by the Huntsman crucible furnace in which a tall chimney drew air through a coke bed and heated a fireclay crucible to a temperature sufficient to melt high-carbon (1 per cent) steel for tools. It used 3 tons of coke to melt 1 ton of steel. Bessemer, who had amassed a sufficient income from a secret process for manufacturing 'gold' paint by cutting bronze into flakes, started in 1854 to try to make steel in larger quantities by melting pig iron with blister steel in a reverberatory furnace. He noticed that when some of the secondary air impinged on pieces of pig iron the outer shells became converted to steel which did not melt, even though the unconverted pig iron in the centre ran out. This gave him the idea of making steel by blowing air through molten pig iron to oxidise the carbon. First he experimented with a blowpipe through the lid of a fireclay crucible in a coke furnace and then with cupola-melted iron in a fixed cylindrical converter with a ring of air nozzles round the base. He found the action must be violent if the conversion was to occur and invented the pear shaped 'Bessemer Converter' which could be tilted both to pour the iron in with the blast holes not under it and for discharge of the finished steel. In 1856 he read his paper 'On the Manufacture of Steel without Fuel' to the BA and sold royalties to a large number of iron works within a month. None of these were able to make good steel by the process and Bessemer had to buy back the royalties. He discovered that he had been lucky in doing his preliminary experiments with a very low-phosphorus, high-manganese iron from Barrow; he only succeeded when he set up his own steelworks and imported Swedish pig iron. In later life he worked on a solar heating furnace and lost a lot of money trying to develop a ship with a saloon on gimbals to avoid seasickness.

William Siemens was similarly able to develop the regenerative fuel fired steelmelting furnace because he had previously (1851) made a small fortune out of a domestic water meter. The idea of a regenerative heat excharger started first with Sterling with a method of heating the air breathed in by tuberculous patients and then to improve the efficiency of the hot air engine with the 'Sterling Cycle' which has the same theoretical efficiency as the Carnot cycle. Siemens tried to develop a regenerative steam engine but failed. and in 1856 he and his brother Frederick applied it to the preheat of combustion air to raise the flame temperature and thermal efficiency of furnaces. In 1861 W. Siemens patented a step-grate gas producer so that a steel melting furnace could be coal fired with the comparable convenience of a gas flame. He used regenerative heaters for both the gas and air and the first successful open hearth steel melting furnace was built in co-operation with the Martin Brothers in 1863. In 1865 Siemens made one ton of cast steel with $1\frac{1}{2}$ tons of slack coal, and in 1868 he set up his own steelworks in South Wales. By 1873 this works was making 1000 tons of steel a week.

Up till about 1950 the Siemens open hearth furnace was rather more in use than the Bessemer because, in fact, the Bessemer process can use only a very small amount of scrap and hence uses more coke in the blast furnace per ton of steel. More recently the top blown oxygen converter has become the most fashionable way of making steel from molten pig iron and the arc furnace for melting scrap. I have been trying to develop a continuous steel melting and refining process for twenty years and Howard Worner in Australia and Glinkov in Russia have done pilot plant experiments on such processes. However the conservatism of the steel industry is at least as great in the twentieth century as it was in the nineteenth.

Maudslay was the man most responsible for the development of machine tools which gave us precision engineering, in place of the earlier lathe with hand-held tools and the crude cannon-boring machines used by Rumford to disprove the caloric theory of heat. After working for Bramah on lockmaking machinery Maudslay set up his own machine shop in 1794, having invented the slide rest for the lathe. He made screws and nuts sufficiently accurately to be interchangeable and he was interested enough in astronomy to make his own telescopes.

THE TRANSITION PERIOD FROM THE NINETEENTH TO THE TWENTIETH CENTURY

In this section we shall consider in detail the inventive careers of two famous inventors who founded new industries that were crucial in the transition from the narrow based technology of the nineteenth century to the affluent society of the third quarter of the twentieth century. Neither of them was the first to think of the particular invention; in both cases it was very much 'in the air' and being worked on by several others. However each of them had the practical ability to grasp what subsidiary inventions and lines of development were necessary to make an idea and then to manufacture a commercially successful series of models. Edison was particularly responsible for the filament electric light and the whole development of the central electricity generating station run by steam engines. Parsons replaced the piston steam engine by the steam turbine and thus opened the way to power units of hundreds of megawatts.

It is rarely possible to gain any clear insight into the thought processes of the great classical inventors because they were usually highly introspective people and did not describe their thought processes even in letters. Their biographers are thus restricted to the realisation of their inventions or to guesswork. Nevertheless a careful study of these biographies can lead to an illustration of many of the principles of invention.

Both Edison and Parsons brought existing ideas to practical fulfilment. The two most important qualities they had were the 'ability to think with the hands as well as the head' and the confidence, dedication and determination to overcome all obstacles. It does not seem to us that they were exceptional in their inventive ability, but rather that they developed the knack of discovering the centre of gravity of the problem and then persisting until they found a solution to this centre of gravity.

T. A. EDISON, 1847–1931[1]

Norbert Wiener suggested that Edison was a transition figure from the crude mechanical inventors of the nineteenth century to the skilled specialists of today with their large-scale systematic experiments. In 1876 Edison established the world's first industrial research laboratory at Menlo Park, New Jersey – now transferred *in toto* to a site outside Detroit where one can even see the cupboard under the stairs where he hid himself for hours on end when he wanted to think.

Edison was noted for his inquisitiveness as a very young child. He was regarded as stupid at his school but his mother took him away and encouraged him to read a great deal. He never learnt to spell but when he was nine he read a book on physical science which caught his imagination and he became self teaching, carrying out all the chemical experiments in the book. At ten he built a telegraph set in the cellar, at twelve he obtained a job as newsboy on the train with the concession to sell food to the passengers. In 1861 he installed his electrical and chemical laboratory in the train baggage car.

At fifteen he read some of Newton's *Principles*, but became baffled, and said later 'it gave me a distaste for mathematics from which I have never recovered'. In another book, Uri's *Arts, Manufactures and Mines*, he found more congenial ideas ridiculing academic philosophers who neglected the steam engine and praising the handcraft operators and artisans who transformed it into an automatic prodigy. He made himself a neat set of telegraph instruments out of junk and became an apprenticed telegraphist. Within five years he became a fast, experienced telegraphist, and devised repeaters which worked well but were not according to the rules. He used to dream inventions in his sleep which vanished when he woke. He was trying to invent a duplex telegraph that would send two messages at once but he lost his job and someone else got there first. Throughout this period as a telegraph operator he was giving much more attention to his experiments (usually electrical) and to reading scientific books than to his work. Faraday's *Experimental Researches in Electricity* with lucid non-mathematical accounts gave him great inspiration.

In 1869 at the age of 21 he left his job with Western Union in Boston and set up as a full-time inventor. He took his first patent in June 1869 for an electric vote recorder for a House of Congress and he managed to persuade various Boston businessmen to advance a few hundred dollars for his work. The vote recorder was rejected as politically undesirable and he decided to invent products that were certain to be in commercial demand. However all his ideas failed and he went to New York penniless; but there he had great good luck and spotted the cause of a disastrous breakdown in the Gold Indicator, a telegraph for gold prices, which led to a good job and the chance to work on his ideas for printing telegraphs. Then he set up with a colleague, Pope, as 'electrical engineers and constructors', left Pope when he found he was doing all the work for one-third of the money, and went back to Western Union as an inventor. Here he received $40000 for a device that brought all the stocktickers in outside offices into alignment with the central one. He set up a small factory employing 50 men to make stocktickers, where he acted as hard-driving foreman as well as inventor.

During the period 1871 to 1876 he worked on the telegraph and its off-shoots, the stocktickers, and he tried to make money for his development work by building and selling machines for which he then had to ensure successful operation. He had endless trouble with the cut-throat financial competition between Jay Goulds Automatic Telegraph Co. and the Western Union and referred to 'small-brained capitalists' as a counter to the 'hare-brained inventor'. When a machine did not work he locked himself and key staff in the works for tens of hours until the bugs were understood and eliminated. On the other hand he would sometimes take all his 'mechanics' off production work to help him with a new idea. Although he was keenly interested in the theory of what was happening (what we call 'physical thinking' in chapter 5) he always used empirical and 'longshelf' methods to give him the clue he needed to solve the problem. He had great intuitive power to know what experiment to do or where to look, endless curiosity and unlimited drive and enthusiasm. He was also a great believer in what is now called 'serendipity', that is both having a whole batch of problems in mind awaiting solutions and also being prepared to use something he came across to solve quite a different problem from the one he started with. In spite of this he could almost 'invent to order', provided the invention required a field of knowledge with which he was already familiar or which he could read about in his highly intensive way. He had a power of 'total recall' for what he read. Although he had no formal scientific education he had made electricity, mechanics and chemistry familiar ground to himself.

By 1874 he had solved the telegraph problem he had been working on for many years, the 'quadruplex' whereby two messages could be sent simultaneously in both directions along a single pair of wires. To help him visualise the highly complicated current flows, as he had little power of abstraction, he constructed a hydraulic analogue – a good example of the fact that one requires a physical picture of what is happening with which to invent. A set of mathematical equations, however accurately they describe the processes, are useless for giving one the creative idea; they can only be used to check it afterwards. The quadruplex really completed his first phase of successful invention, but he also constructed a battery-operated arc lamp and made some observations on the production of sparks by high-frequency electromagnetic waves which he called 'a true unknown force'.

In 1876 he built his own laboratory in a small village 25 miles south west of New York and abandoned manufacturing to concentrate on the 'invention business'. Menlo Park was staffed at first with fifteen associates – all skilled mechanics, but later he employed a mathematician, a glass blower and many · others. This was probably the first specially set up industrial laboratory in the world. At this period pure scientists such as Clerk Maxwell had a great contempt for the practical engineers (as indeed many still have) but Edison was frankly commercial, called himself an 'industrial scientist' and was equally contemptuous of the mathematicians and physicists.

In 1875 Edison was asked by Western Union to work on the telephone; they rejected Bell's work because he wanted $100000 for the patents. They were reluctant to improve their telegraphs on the principle enunciated by Andrew Carnegie that 'pioneering don't pay'. In 1912 Mr Justice D. Brandeis remarked that 'the great organisations are constantly unprogressive. They will not take on the big thing. Take the gas companies, they would not touch the electric light. Take the telegraph companies, the Western Union, they would not touch the telephone.'

Edison's contribution to the telephone was to separate the microphone from the earphone and by 1878 he produced a really successful carbon microphone in which the sound waves varied the resistance. Western Union sold out their telephone interests to the Bell Company in 1879.

Edison believed in the method of serendipity, starting in one direction but observing a phenomenon which, to his well-stocked mind, suggested a solution to another problem. Thus his work on the gramophone arose from a morse telegraph repeater he had made earlier which embossed a paper disc, together with his observation that the telephone receiver diaphragm vibrated with considerable amplitude. In 1877 he produced his crude but effective phonograph with a tin-foil cylindrical sheet; it was rotated by hand with a flywheel to give an even speed and only recorded for one minute. This brought him world-wide fame and Edison predicted all the uses to which tape recorders are now put and devised a scheme for making many wax copies with a master matrix, but in 1878 he abandoned all work on it for ten years and started work on his most important invention of all.

In September 1878 he realised that the arc light could never be subdivided and used for domestic lighting as gas was used. He told a reporter of his vision of a central lighting station providing electric light for the whole of New York and said he would solve the problem in six weeks.

He realised that he must have a subdivided circuit with many small, about 8 cp, lights in parallel each with its own switch, and that these lights must have a high resistance of 100–200 ohms. His objective, he told a reporter, was not making his fortune but 'getting ahead of the other fellows'.

First he managed to keep a filament of carbonised paper alight in a vacuum for about eight minutes; then he worked with platinum filaments and by using a regulator that shorted the filament when it got too hot kept it alight for ten minutes, telling reporters that his success was already assured. Silvanus P. Thompson published a paper saying that anyone who tried to invent an incandescent electric light was doomed to failure and that Edison's talk of subdivision of currents showed the most airy ignorance of the fundamental principles of both electricity and dynamics. 'Edison got himself into trouble purposely by premature publication so that he would have a full incentive to get himself out of trouble' said his mathematician F. R. Upton. When Upton arrived at Menlo Park, Edison asked him to work out the volume of a light bulb. After more than an hour of calculation Edison measured it with water and a measuring cylinder in a few seconds.

Edison set up the Edison Electric Light Company and raised $50000 for his research, and the gas lighting companies became alarmed. In Britain a Parliamentary Committee (with Lord Kelvin and John Tyndall's agreement) reported that Edison's projects were 'unworthy of the attention of practical or scientific men', while Sir William Preece in a Royal Institution lecture said 'in a circuit where the emf is constant and we insert additional lamps, then when these lamps are joined in one circuit, i.e. in series, the light varies inversely as the square of the number, and in multiple are as the cube of the number. Hence a subdivision of electric light is an absolute *ignis fatuus*.'

In fact Edison 'intuitively' understood Ohm's law better than any of the professors and realised that only by producing a lamp with 100–200 ohms resistance and developing a constant voltage dynamo could he produce his light distribution system without using too much copper. In January 1879 he produced his first high resistance platinum lamp and then tried platinum-iridium, boron, chromium, molybdenum, osmium, and nickel; he could not work tungsten so did not try this. He made great efforts to obtain higher vacua and discovered that if he heated the filament while bringing the pressure down he could drive off occluded gases and raise the melting point a little. However his financial backers had grave doubts when he went to them for more money because this line had brought him up 'a granite wall a hundred feet high'. He abandoned platinum and his team tested 1600 different filament materials.

By the middle of 1879 he had produced a direct drive 350 rev/min dynamo giving a nearly constant voltage and of much higher efficiency than previous ones (90 per cent) because it used laminated sheet iron, a lot of iron and heavy wires. He had obtained a technique for sealing the bulbs with 10^{-5} atm.

Then he returned to carbon using his new technique of degassing. He produced a 200-ohm filament, 6-in long 1/64-in diameter from carbon black and tar, carbonised. On 21 October 1879 his team obtained $13\frac{1}{2}$ h life with a carbonised cotton thread; this was followed by all kinds of other carbonised vegetable fibres and in December after fourteen months of constant experiment they obtained 170 h with a carbonised cardboard filament. This success was published in the press and Edison was able to get more money. Finally after many more trials carbonised bamboo splinters were used and a life of 1200 h was obtained.

Edison then embarked on three years of development of the whole system of central station lighting. He developed his dynamo with a voltage regulator, fuses, switches, the screw-in lamp socket, underground mains with tar insulation, the feeder and main and three-wire circuits to reduce copper requirements for a given number of lamps. In 1880 he installed a complete electric lighting system on the SS Columbia: he also built and successfully operated an electric railway in that year but ran up against patent problems and dropped it. By Christmas 1880 he had installed a complete pilot plant lighting system at Menlo Park with 425, 16 cp lamps and eight miles of wire, operated by a 120 hp generator. This demonstration was completely successful and in spite of the gas companies' lobbyists, Edison was granted a franchise to install a central lighting system in New York. Edison did a careful economy trial on his installation and a market survey and concluded he could operate a 16 cp lamp on 0.4 lb coal/h giving the same cost as a gaslight of equal power. He then went back into the manufacturing business and set up companies which built all the components and installed the complete system.

The Pearl Street System in New York was started in September 1882 but there were many problems in getting it to operate and then setting up factories for all the components, which kept Edison fully occupied for many years. During this period many central generating stations were set up and the foundations of the whole electricity industry laid down. Edison said he was not inventing but managing during this period and he had many patent disputes as rival groups were always able to find someone who had tried a similar idea before Edison. However Edison's group won the final action because he was the first person to realise the need for a high resistance filament and to make a practicable system. In court in 1890 he remarked 'I have had a great many mathematicians employed by me for the last 10 years and they have all been dead failures', 'The Mathematics always seemed to come after the experiments not before', 'I can hire mathematicians but they can't hire me'.

During this period he discovered the Edison Effect – the conduction of electricity from a negative heated filament to a plate in a high vacuum which he attempted to use as a filament temperature controller but which later led to the discovery of the electron, the valve and the whole subject of electronics. (If he had had a scientific training, perhaps Edison would have followed this up himself.) Between 1889 and 1904 he developed the 35 mm ciné camera to a practical stage.

Around 1889 when his rival Westinghouse was beginning to develop an a.c. distribution system so that high voltages could be transmitted over long distances, Edison resisted it fiercely.

In 1887–8 he returned to the phonograph and developed a clockspring drive and speed regulator, producing a commercial device. In 1892 he sold out of his General Electric Company and then spent several years trying to produce an iron ore concentrate but when the Mesabi range of high grade iron ore was discovered he realised this must be abandoned with a loss of $2m of his own money.

His last great work was the development of the nickel–iron storage battery which occupied a team of ninety people under his direction from 1900–1909.

CHARLES PARSONS[2]

We can gain some insight into the mind of Charles Parsons from his writings and he provides an example from which much can be learnt. His father, the Earl of Rosse, was of an inventive turn of mind and constructed, at Birr Castle where Charles Parsons spent most of his boyhood, being educated privately, a forge and workshop capable of making any kind of machinery and especially astronomical telescopes. With his father and brothers he constructed steam engines and electromagnets as well as telescopes; they went to sea in an iron sailing yacht. There was no Engineering School at Cambridge then but Parsons passed eleventh in Mathematics in 1877 being good at problems but weak at bookwork. His rooms had a litter of engineering models, including an epicycloidal steam engine for very fast rotation, but he appears to have used only simple arithmetic and orders of magnitude calculations to test his ideas rather than any advanced mathematics. He said later that the five years spent studying mathematics were the most severe years of strain in his life. After Cambridge he had three years as a Premium Apprentice with Sir William Armstrong at Newcastle where he learnt much about mechanical engineering research and construction. Parsons had already decided that he wanted to achieve practical developments rather than follow the older belief in concentrating on speculative natural philosophy. The ideas of thermal and mechanical efficiency and the need for a large expansion ratio in the steam already existed and Parsons could see quite clearly the limitations of the single expansion steam cylinder engine. In 1884 he joined Clarke Chapman at Gateshead as a junior partner and they decided to work on electric lighting generators for ships (arc lighting at that time) so Parsons abandoned his work on rockets and torpedoes, and he says he then decided to try to build a steam turbine which could directly drive a very high speed generator. In 1885 he built one which used 150 lb steam per kWh. This gave 6 hp at 18 000 rev/min. By 1889 when he decided to leave Clarke Chapman he had 20 hp turbines using 63 lb steam/kWh.

Both the impulse and the reaction turbine were old ideas and de Laval was working at the same time in Stockholm on the Heron-type reaction turbine in which steam emerges at high velocity from S-shaped arms like a catherine wheel. Soon afterwards de Laval produced the impulse wheel with several convergent-divergent fixed steam nozzles sending supersonic jets of steam at the vanes. The advantages and the theory of this type of nozzle were understood by 1888. De Laval also introduced the idea of a flexible shaft which was self-balancing at the very high speeds necessarily involved in the impulse turbine where the blade speed ideally equals half the steam speed, and these turbines have been constructed up to 500 BHP.

Parsons also used the flexible shaft, but had the brilliant idea of expanding the steam in several stages so that the velocities did not have to reach the high supersonic levels (several thousand ft/s) of the single stage expansion; this enabled very much bigger turbines to be built, now up to many hundred MW, without exceeding the strength of steel blades by centrifugal force. The steam flows steadily through the turbine from one stage to the next without using valves or reciprocating parts. At first Parsons' turbines used more steam than piston engines and were only valuable because the turbogenerator was much more compact and smoother running and enabled a very high-speed d.c. generator to be coupled without gearing. His real aim however was to economise on fuel and by 1897 he had worked out near optimum principles of design for the shape and spacing of the blades. Parsons obtained one thousand-fold expansion of the steam without intermediate losses whereas the triple expansion piston engine can only achieve 16 per cent thermal efficiency.

Already in this period he had invented the parallel flow turbine in which the high pressure steam entered at the centre and flowed both ways outwards to balance the end thrust, and he had realised that the blades should get longer as the steam expanded. He had experimented with turbines in which the stages were concentric on a single disc but appreciated the disadvantages compared with axial flow.

At the same time he had to redesign the generator to take advantage of the high speeds – he used a low field magnet of cast iron to reduce hysteresis losses in the armature and he produced a voltage regulator which controlled the field. In 1885–9 nearly 400 turbogenerators producing up to 75 kW of very low voltage, heavy current were installed in naval ships all over the world and by 1885 the steam consumption was down to 34.5 lb/kWh.

In 1889 he decided to dissolve his partnership with Clarke Chapman and set up his own firm, but by the agreement all patents taken while Parsons was a partner were the property of the firm. For three years at arbitration proceedings Parsons tried to prove his patents had very little value (only £2000) while Clarke Chapman tried to value them highly. Parsons' experts tried to prove he had been forestalled by producing a patent of 1848 by Robert Wilson that already had the idea of a number of rotary engines on a common shaft worked by the same steam in succession. The expert for Clarke Chapman (Sir William Thomson) claimed that Parsons had evolved the idea and made it work successfully by achieving a good method of lubrication, the semi-elastic bearing to give dynamic balance, the parallel flow system and the use of fixed blades of similar curved shapes to redirect and accelerate the steam for the next rotary blades.

The arbitration was closed by Clarke Chapman, declaring that to show his faith in the patents he would use them without Parsons. Parsons worked on a radial flow turbine for a few years but had many troubles with it and in 1894 he was able to buy back his patents from Clarke Chapman. It is recorded by Parsons' patent agent Dugald Clerk that when Parsons described a new idea it was not till after he had spoken for twenty minutes that his meaning dawned on the agent; he clearly did not have the power of expressing the essence of his ideas in a few words. It seems clear that his great contribution was to recognise the physical problems such as lubrication, steam leakage and condensation into drops, vibration, strength and blade shape and study them deeply until he had solved the problem of their development, either by an ingenious new idea or by a design based on physical understanding of the process. The basic ideas of the turbine were not new, but by means of a series of subsidiary inventions he made it work and then brought its steam consumption steadily down until it gave an overall efficiency far in excess of the triple expansion piston engine.

When Parsons applied in 1898 for an extension of his 1884 patent it was concluded that his expenditure to that date had exceeded gross profits after allowing 7% interest on capital; thus 14 years' work had still not produced any profits. It was only Parsons' personal fortune which had enabled him to continue against tremendous apathy and hostility and to carry the turbine beyond the electric lighting stage to the propulsion of the *Turbinia* at 35 knots in 1894. Parsons was granted a five year extension of the 1884 Patent because the House of Lords' Judicial Committee considered he had not been adequately remunerated. They considered that if an applicant had not cleared £10000 then he should have an extension.

In order to prove his ideas about the advantages of the turbine for ship propulsion, Parsons set up in 1894 the Marine Steam Turbine Co. with £25000 share capital, and built the 100-ft *Turbinia* with a 1000 hp radial flow turbine. After the success of this experimental ship ($32\frac{1}{4}$ knots on measured mile) in 1897 a new Company, the Parsons Marine Steam Turbine Co., with initial capital of £240000, was raised by·a shares issue.

To design the *Turbinia*, Parsons carried out model experiments on hull design (using 2 ft and 6 ft models) for high speeds, being especially concerned with the question of bow lift and the effective horse power required. He studied the latter with a rubber cord driven propeller doing 8000 rev/min and predicted the power required to drive the *Turbinia* within 3 per cent from his model studies. The propeller design was studied on the full scale, and seven different types were tried on thirty-one trials at sea before he obtained the best results with three propellers on one shaft. He used a torque measuring spring coupling which led him to fit three propeller shafts, and he also changed to parallel flow turbines, one to each shaft before he got the speed up to 34 knots. My father was a junior naval officer on one of the ships at the 1897 Spithead Review when Parsons made rings round the Navy. Nearly eight years later (March 1905) the Navy decided to adopt turbine propulsion.

One other anecdote of a later period shows Parsons in a different light. Lord Fisher wrote to Parsons on 3rd November 1912 '. . . wish you could see your way to the continuance of the Turbine in connection with the Internal Combustion Principle'. On 21st November 1912 Parsons said at the Royal Commission on Fuel and Engines 'I do not think the internal combustion turbine will ever come in. The internal combustion turbine is an absolute impossibility'.

It was left to Whittle and others twenty-five years later to prove that one could design an air compressor so efficient that it did not absorb all the power of the turbine.

THE TWENTIETH CENTURY

A very careful analysis of the twentieth century and many of its chief inventors has been made by Jewkes et al.[3] They conclude that corporations as opposed to private inventors are certainly now accounting for a larger proportion of useful inventions than formerly. The figures taken from the patent statistics are:

The Percentage of Patents Taken by Corporations

	1900	1913	1936/8	Since World War II
U.S.	18		58	54–64
G.B.		15	58	68 (1955)

Jewkes believes that these statistics exaggerate the corporate proportion because (1) corporations tend to patent a large number of inventions of subsidiary importance or for competitor blocking, whereas in some fields such as oil refining and photography, the really big new ideas come from individuals; (2) corporations may be assigned patents by individual inventors at an early stage. The apparent dominance of corporations is greater in some fields such as chemicals and electronics and less in others such as aeronautics and internal combustion engines. When a subject has developed to the point where it requires a highly trained man working full time and backed by many hundreds of thousands of pounds worth of development before a new idea can be tested it is no longer possible for the individual to succeed alone as most of the nineteenth century inventors succeeded. However it is just as true now as it was then that 'the Establishment', which now includes the big corporations, has a very strong inevitable opposition to any major innovation. The main reasons for this are that experts see their expertise as being made redundant, and that managers have to make a big effort to be able to manage something quite different. The classical three-stage approach to innovation:

(1) It will not work
(2) If it does work it will not be economical
(3) I thought of it first

has even less chance of reaching stage 3 now than it did in the nineteenth century.

A more serious criticism of the twentieth century is that the education in fundamental principles of physics and chemistry, which is so essential to any inventor in most fields, tends to be associated with a discouragement by the teachers of the would-be inventor. It is one of the main aims of this book to provide a weapon against this discouragement.

Jewkes gives some examples of successful twentieth century inventors and there are many others who were well trained in the fundamentals of their subjects and started their development work without the resources of a large corporation. These inventors include Whittle (aircraft gas turbine), Land (polaroid camera), Wankel (rotary piston ICE), Moulton (hydrolastic suspension and small wheel bicycle), Cockerell (air cushion vehicle), Ferguson (tractor with integrally mounted plough) and Buckminster Fuller (geodesic domes). We have not space to discuss the twentieth century in detail but the conclusions of Jewkes et al. and their case studies have been taken into account in the next three chapters.

REFERENCES

1. Josephson, M., *Edison: A Biography*, Eyre and Spottiswoode (1961)
2. Appleyard, R., *Charles Parsons: His Life and Work*, Constable (1933)
3. Jewkes, J., Sawers, W. and Stillern, R., *The Sources of Invention*, Macmillan (1969)

3 WHAT NEEDS INVENTING?

M. W. Thring

HAVE ALL THE WORTHWHILE INVENTIONS BEEN MADE?

The nineteenth century was a great period for the individual inventor. Siemens and Bessemer developed processes that led to steelworks producing hundreds of thousands and, eventually, millions of tons of steel per year; Edison and Parsons founded great companies, while the inventors of small things like the safety pin or the bicycle free-wheel made their fortunes. Hundreds of other inventions, however, fell by the wayside either because they were not economic, the materials were not available or they simply wouldn't work—like steam operated helicopters which were no more practical than Leonardo da Vinci's human power helicopter designed five hundred years earlier.

This period was, however, the great period of Samuel Smiles' 'self help' and indeed he wrote several books about inventions in this vein. The only criterion of success was whether the inventor made a fortune, and a fortune from machine guns was as good as a fortune from safety pins.

It is often said that the day of the private inventor is over and that only the vast corporations can afford to develop new ideas. This is partly true as long as invention is judged purely by the profit motive: even then it is only partly true since in America a number of Professors of Engineering have invented new devices, particularly in electronics and instrumentation, and have successfully formed their own small businesses to manufacture them. Many of these only succeed as a result of large Government contracts, but the Xerox process and the Land Instantaneous Camera are examples where individual inventions have led to large new successful business corporations.

When Neville Shute was a works manager in the 1930s, it was still possible for a new firm to produce a viable new aircraft from scratch in two or three years and sell it, as he describes in his autobiography, *Slide Rule*. Now even a vast corporation cannot develop a new engine or a new airframe without direct or indirect government subsidy to cover most of the development costs.

What then is the possible role of the private inventor? Clearly he cannot compete with the vast corporation unless he gets substantial Government backing (e.g. the Hovertrain *vs.* British Rail). However, the vast corporation is in the process of strangling itself by its single minded pursuit of profits, at the expense of the damage which is being caused at present, and will be caused in the future, to people and to the environment in the fifteen ways shown in table 3.1. The private inventor has the fundamental advantage that he can choose his objective freely without being caught in the trammels of the dividend. It is true that this makes it extremely difficult for him to get

Table 3.1   The harmful results of technology

GROUP 1:   DAMAGE TO THE ENVIRONMENT

I   *Air pollution.* CO; unburnt hydrocarbons; soot; Pb compounds; SO_2; NO_x; HCl; HF; HCN; radioactive gases and dusts; CaO

II   *Water pollution.* Factory effluents including sulphite lye; salts of Pb, Cd, Hg; sewage; fertilisers and pesticides washed from farms; oil; thermal pollution

III   *Land pollution.* Drums of poisonous waste; litter; slagheaps; derelict factories; radioactive and Pb fallout

IV   *Noise and vibration.* Aeroplanes; cars and lorries; domestic and factory machines; mains hum

GROUP 2:   DAMAGE TO PRESENT HUMANITY

V   *War machines.* Bombs; warplanes; ICBMs; guns; tanks; defoliants

VI   *Poverty — underdeveloped countries.* Slums and poor farms in rich countries

VII   *Accidents.* Aeroplanes; cars; ships; mines; factory; home

VIII   *Unnatural and unhealthy life conditions, especially in cities.* Overcrowding; ugliness; tasteless food; traffic jams; loneliness; lack of privacy; lack of exercise; high-rise flats

GROUP 3:   DAMAGE TO FUTURE GENERATIONS

IX   *Fossil fuel depletion.* Oil; natural gas; coal; concentrated U ore

X   *Radioactive waste.* From nuclear power-stations; lack of permanent safe disposal

XI   *Metal ore depletion.* Mesabi iron ore; Cu, Ni, Cr, Zn, Hg, Sn, W

XII   *Misuse of land.* Monocropping; hedge removal; burning stubble; lack of humus-making fertilisers; concrete and asphalt; compaction by heavy machinery

XIII   *Misuse of fresh water.* Artesian wells lowering water level; dams that fill with silt; water waste in cooling towers, lavatories, etc.

GROUP 4:   HUMAN PAID WORK

XIV   *Under-employment of talents.* Disappearance of craftsmanship; machine feeding and machine minding; assembly-belt work; no connection with quality of product

XV   *Unemployment.* Caused by productivity rise above point where it can be absorbed by living-standard rise

money for his inventions, but the inertia of big industry makes it almost equally difficult to get money for any other invention. In any case there can be little doubt that the aggregation of these problems will necessitate a complete change of direction for industry which will have to start some time in the next fifteen years. Thus the private inventor can choose to work on those inventions which essentially contribute to the solution of these problems. In this way he can be sure that his inventions are valuable to humanity, which is the best way to ensure that one has the emotional drive and persistence to succeed in invention. He can also find plenty of new things to invent because industry does not explore this field.

THE MORALITY OF MACHINES

The kinds of question we have to consider are:

Is it better to invent a faster plane or a safer one? A plane for first class business travel or one for cheap popular travel?
Is it better to increase car acceleration or reduce fuel consumption and air pollution?
Is it better to design a folding pram or a tank? Under what circumstances if any is it humane to invent improved weapons of war?
Is the development of an efficient breeder reactor more important than finding a safe way of disposing permanently of fission products? Is nuclear power capable of being extended to serve all mankind in the next century? How much inventiveness should be given to the problems of the under-developed countries, and to the needs of humanity in the twenty-first century?
Will robots increase unemployment? and should we make them?
In what direction should railways improve, e.g. speed, reliability, fuel consumption and type of fuel, noise, safety, goods carrying capacity, frequency of service?
Is it desirable to design machines with a lifetime of only a few years?

To answer moral questions one must make basic assumptions about human needs. I shall make two such assumptions which are essentially a matter of belief and cannot be 'proved' on purely logical grounds.

(1) What human beings require from life is essentially the subjective self-fulfilment by which one measures the quality of life. The accumulation of physical possessions and the consumption of food and other life necessities are only of value in so far as they contribute to the quality of life, and if they are acquired and consumed to the point where they cause a deterioration of the quality of life or the health of the individual their true value to the individual declines. Just as a gradual increase in the daily calorie intake causes at first a rise and then a decline in individual health as one overeats, so a gradual increase in the monetary standard of living causes at first a rise in the quality of life (as the problems of shortage of necessities, inadequate education, travel and human relations are overcome) and then, as one rises beyond all the requirements of a full life into the region of affluence, a fall in the quality of life for a number of reasons. This is demonstrated in the most affluent societies by the rise in the number of people opting out of normal life by suicide, alcoholism, drug addiction and formation of isolated communities, by the increasing incidence of crimes of violence against the individual and vandalism, and of stress illnesses and nervous breakdowns. There are many reasons for this fall in the quality of life or self-fulfilment when affluence is excessive. Because the sum total of goods available is limited by world resources, an individual who consumes many times his

33

share of the cake either comes to regard his wealth as the justification of his life (i.e. he transfers the idea of fulfilment of himself to the idea of possessions as the aim of life) or else he has an uneasy conscience when he realises how wasteful he is of resources, while people elsewhere in the world are starving. Another reason is that the possession of great wealth causes many human relations to become soured by the fear that people might be friendly for what they can gain.

(2) Self-fulfilment is essentially achieved by using all one's talents and potentialities to the greatest possible extent to produce something which is of value to other people. This means that one must consider the consequences of one's inventions in the context of their effect on all humanity (present and future) and on the whole environment in which we live.

Using these two postulates one can put all kinds of possible inventions, and consequences of inventions, on a moral scale. In the highest category come all those inventions directly concerned with and aimed at increasing the potential of the individual for self-fulfilment. This includes: all the help the inventor can give to the medical profession in restoring people to normal health and activity; improvements in education, communication, and leisure travel; the invention of new devices for artistic and craft activities.

Immediately below the individual-extending category come all the inventions which enable humanity to earn the necessities of a full life without having to endure excessively long hours or highly repetitive work.

In the positively evil category at the bottom come all the inventions of war and torture, and all the harmful consequences of cheap, careless or wrongly-motivated inventions such as

Pollution
Noise
Ugliness
Exhaustion of minerals and fuel.

It is a sad commentary on our society that in the last two hundred years the inventor has increased the productivity of the soldier (measured by destruction produced by one hour's work) by a factor of more than ten thousand and has increased chemical pollution and noise by a similar factor, while in the same period he has increased the productivity of the factory worker and the farm labourer only by a factor nearer ten than one hundred.

Just above the line that separates good from evil come the inventions in the category we can call cosmetic engineering or fashion engineering. These are the inventions that are not harmful in themselves but are inessential to the optimum standard of living corresponding to the highest quality of life. They are thus the causes of the four harmful consequences of wrongly motivated engineering. In this category comes built-in obsolescence, rejection of little-used objects because of fashion changes, buying unnecessary gadgets, inventions for advertising, unnecessarily large machines (for example cars), gimmickry, throw-away containers, packages and many electronic devices.

In my two books, *Man, Machines and Tomorrow* and *Machines: Masters or Slaves of Man?*, I have developed the thesis that a continuation of our present line of development of the 'affluent society' will lead inevitably to a disastrous fall in the quality of life in the developed countries, as well as a complete failure to raise the standard of living in the underdeveloped countries to a reasonable level. Many other people have reached the same conclusion that a society based on the idea of a perpetually rising standard of living must destroy all that is really worthwhile for its members. I have concluded, however, that this fate can only be averted by a complete change of the ethos of our society, from that in which we judge success in life by possessions to one in which we would judge success entirely by self-fulfilment. One can envisage a society in the twenty-first century in which man lives in near-equilibrium with his environment, with the minimum use of raw materials by fuel economy, complete recycling of all metals, no throw-away goods, all consumer goods built to last many decades, and near zero pollution. I have called this the 'creative society' because the excitement of life would come not from the acquisition and variety of the possessions of the individual but from the creative self-fulfilment of the arts and crafts. Everyone would have leisure and energy to be creative because the machines would relieve them of all drudgery, and enable them to earn their living using much less of their daily energy than we use now even in the rich countries.

In the remaining sections of this chapter we will discuss the various inventions necessary to avert the disaster of the advanced affluent society and enable us to move into the twenty-first century with a possibility of all human beings finding their own self-fulfilment.

TECHNOLOGY FOR THE LESS DEVELOPED COUNTRIES

Schumacher has coined the expression 'Intermediate Technology' for the idea of producing a special technology for the less developed countries intermediate between that of the highly developed countries and their present technology. It is certainly useful to have a name for the technology that needs inventing, but I query whether it really has to be in the direction of our technology at all, since this implies that the poorer countries will have to have a proportion of all the bad things which are associated with our technology. For this reason I prefer to use the term 'Humane Technology' for the goal which we have to aim for in both the rich and poor countries.

The most important problems which must be solved in Humane Technology are to provide everyone in the less developed countries with enough good food, education, travel, housing, clothing, without creating:

(1) Pollution to a health-damaging degree
(2) Excessive destruction of wildlife and wilderness
(3) Unemployment

(4) Too rapid depletion of the world's fossil fuel resources, particularly oil on which so much of our present transport system is based.

At the Stockholm Conference in 1972 the attitude of the less developed countries tended to be 'We want the benefits of the "affluent society" too and we don't care if we pollute as much as the rich countries in getting them'. However, if the inventors can find ways of producing the real benefits without the 'disbenefits' which destroy the quality of life, then the undeveloped nations are sensible enough to prefer such an alternative technology. The only way to do this is for the inventors to visit these countries and apply their inventive skills to solve the problems they find. The suggestions given here are based only on a few short visits to Nigeria and to Argentina (which is a country with a technology already intermediate between that of the rich and poor countries, as shown by the fact that the energy consumption per head is close to the world average of two tons coal equivalent per annum).

There are many problems connected with water for domestic and agricultural purposes. In many places there is a brief monsoon period in which all the year's rain falls, and the problem is to store it for use in the dry season. Unfortunately large dams can only be built where the rain falls on suitable mountains, and in any case stop the valuable silt from being carried downwards; such dams would become silted up within a couple of generations. The invention of a way of storing the water in small reservoirs all along the river, or a way of allowing the silt to by-pass the big reservoir, could clearly be of great value. Evaporation must be reduced to a reasonably low proportion of the amount of water stored.

The Intermediate Technology Development Group[1] has already produced a family-sized water-storage unit. This consists of a collection of heavy plastic sheets and sacks: the family dig a deep cylindrical tank in the ground, lining it with plastic sacks filled with earth. A conical collector sunk into the ground around the tank top enables the tank to be filled with rain water during the rainy season and a white plastic cover stretched across the surface reduces evaporation in the hot period.

Another important water problem is that of lifting water from a well, or low-level channel, to a level at which it can be used to irrigate land. The use of a diesel-engine or electric pump is the high level technology solution, but the problem is to use cheaper energy, such as combustible refuse or solar energy, and a low capital cost system. Professor Dunn is studying the possibibility of the Humphrey pump fuelled with sewage methane, while very low-efficiency solar powered systems are being studied at Ahmadu Bello University in Nigeria.

Finally we have the problem of irrigating deserts to grow crops. The only practicable source of water for this purpose is sea water, which is so saline that it must be distilled for use, and probably the only energy source cheap enough for this purpose is solar energy. The basic problem is to find a low capital method of condensing the water vapour and applying it where it is wanted. It is not difficult to design a feasible solution but it will not be developed by technologists until food and energy in the rich countries become relatively much more valuable commodities than they are at the present time.

In the case of farming, the whole emphasis in the rich countries has been to get as much crop from a given land area as possible, but also to do this with as little labour as possible. This last requirement does not apply to the less developed countries. Thus the farm machines of the rich countries and their methods of using large quantities of nitrogen fixed from the air at great cost in fuel, and of potassium, phosphorus from exhaustible deposits, which are eventually run off into the sea, are quite unsuitable for the developing countries. There, methods of recycling N, P and K from compost and city night-soil without disease spreading, and the more extensive use of crops that bind their own nitrogen from the air are much more relevant, and the mistake of damaging the soil for maximum short-term profit (by cutting down hedges, mono-cropping, panning with heavy tractors and extensive use of pesticides) would be fatal. The main problem is how to cultivate the soil to produce the best crops either with hand tools or local-fuel-powered machinery which is low in first cost and easy to maintain and does not damage the soil by panning. Alternatives to ploughing which give equally good or better weed destruction and soil conditions (especially permeability to air and water) and use much less power have been invented, but their lack of modification to suit various soils and climates leaves plenty of scope for valuable inventions.

The leaf fractionation process, developed especially by N. W. Pirie, can produce two tons of human food protein per hectare-ann. in the English climate and up to six tons in a tropical climate, provided there is plenty of water. This process extracts protein from any young fresh leaves – even fresh water weeds – by the same process that a ruminant (cow, sheep or goat) uses, of rupturing the cells, squeezing the juice out and then extracting the protein from the juice by curdling it by acid or heating and then separating the curd. While the machinery necessary to work the process, on a 10 hp scale continuously and 1 hp as a batch process, has been built, there is tremendous scope for inventive improvements, and especially for developing a very efficient and cheap one manpower unit for protein production in remote villages. The only commercial machinery used so far has been adapted from sugar-cane crushing machinery or screw extrusion presses developed for processes requiring entirely different mechanical treatment of totally different materials.

Much experimental work has been done on sewage and agricultural waste-fermentation to produce methane or methyl alcohol and compost, to recycle human and animal sewage and waste materials as sterile compost and to conserve P, K and trace elements. Crops such as legumes, clover and alfalfa which can bind their own nitrogen using solar energy can avoid the high energy consumption required for chemical nitrogen fertilisers. However, there is still room for inventive work in all these fields.

There are also many problems in the field of low-cost buildings for crop storage and food preservation which require inventive solutions.

Finally there is the problem of transporting agricultural produce, tools and machinery across country where there are no roads and the terrain is difficult. The helicopter and hovercraft can be used, but these require far too much fuel; the inventor must find something that uses no more fuel than a lorry on a road. Airships will undoubtedly provide one worthwhile solution, especially if the boundary layer propulsion device (see chapter 10) can be applied to reduce airship propulsion power still further. A 'mechanical elephant' walking on spring legs (see 'Centipede' in chapter 10) is another. A modification of the centipede with soft pneumatic legs like giant sausages is being designed to carry tree trunks across country where the protruding tree stumps can be as much as two feet high.

DANGEROUS WORK

While there will always be people who enjoy doing dangerous things like rock climbing, caving and sailing round the world alone, it is not right to persuade people to do a dangerous job of work if a machine can do it better. Obviously if a dangerous process can be done by a true robot (i.e. without real time human control) then this is a relatively straightforward use for robots. There are, however, many other cases where the operation is too subtle for a pre-programmed robot and requires too much adaptation for unforeseen circumstances, so that a skilled human must be controlling the robot all the time it works, or at least be capable of intervening in its normal movements and, therefore, be fully informed of what it is doing and receiving sense impressions from it. The Russians have operated a telechiric robot on the moon controlled by radio from Earth with television visual feedback, so the examples given below require straightforward inventiveness for their solution rather than the development of any new scientific techniques.

The most obvious problem to be solved in this field is the mining for the solid minerals (such as metal ores, gold, diamonds, coal, shale and tar sands) in such a way that they can be brought to the surface without men ever going underground, just as one brings oil, natural gas or water and soluble salts. It is possible to gasify coal to produce gas *in situ* and many experiments have been done on this, but it is very difficult to make good producer gas, even in a specially designed gas producer fed with sized coal, so it is certainly not an efficient way of using limited coal resources. I have shown by calculations from the published gas analysis that most of the coal is left as coke underground, and there are grave problems of heat absorption by water evaporation, gas leakage into surrounding fissured strata, and moving the pipes to gasify a new region once the system has been ignited. A real possibility is, however, a surface controlled, mobile gas-turbine electricity generator mole, which receives air by a pipe from the surface, compresses it, burns coal in a high pressure region in front of it, expands the gases in the turbine which drives the compressor and generator and exports the electricity and waste gases to the surface. Another possibility is the solvent extraction of the coal with a suitable liquid such as anthracene.

Similarly there are possible processes for extracting the hydrocarbons from oil shale and tar sands *in situ* without mining the solid residues by the use of heat or steam. However, to obtain metal ores such as copper and uranium which often occur in thin seams deep underground, or solid coal at the surface (needed, for example, for coke making), then a 'telechiric' (distant-hand, Gk.) mole system which can locate the seam, crush the mineral and convey it to the surface is essential. This makes sense not only from the point of view of saving men from working in dangerous situations and perhaps contracting pneumoconiosis and lung cancer, but also from the straight economic point of view. It enables us to mine minerals which are quite unreachable by men because they are too deep under the ground, far under the sea, in very steeply sloping seams or in very thin seams less than half a metre thick. Twenty years ago the proven coal reserves of Britain were calculated to be enough to provide us with two hundred million tons a year for two hundred and fifty years. Now the Coal Board says we have enough to provide us (by conventional mining techniques) with one hundred and twenty million tons a year for one hundred years in spite of the fact that during this twenty years they have discovered more reserves than they have used coal. The reason is that the relative costs of human mining have made three quarters of our coal reserves uneconomic so that the invention of a surface controlled machine which enables coal to be found, cut and brought to the surface without any man ever going underground would provide us with all the energy necessary for this country, if we are economical, for two hundred years. It could be done by a self-propelled tunnel boring mole which crushes the coal and pumps it to the surface in water or by a modification of conventional long wall mining methods and machines to enable surface controlled telechiric miners to do all the tasks that men do at present.

Telechiric machines have already been developed to work at the bottom of the sea but here again there is great scope for inventive improvements to enable men on shore or in a surface ship to drill oil wells or mineral mines on the bottom of the continental shelf, to pick up nodules from the sea bottom, to 'farm' the continental shelf, and to repair cables, make biological studies and prospect at the bottom of the deepest oceans.

Another group of problems in this field are those of extinguishing unintentional fires and rescuing people from dangerous situations. To rescue people from burning buildings and crashed aircraft, probably a powered suit of armour with water cooling and thermal insulation or a man-carrying chariot which can climb stairs and force its way into the fire will have to be invented. For extinguishing burning buildings in which people are not at risk, a telechiric fire extinguisher which can be controlled from outside to move into the heart of the fire and apply the most effective extinguishing agent is required. Another problem is to develop a telechiric device which can extinguish an oil-well fire and cap the well by remote control.

MEDICAL ENGINEERING

The humane inventor gives a very high priority to the invention of machines to help doctors and surgeons in their health work, to reduce the incidence of disease, accidents and disablement, and to help elderly, crippled, blind and otherwise handicapped people to lead a more normal life.

Medical engineering needs innovative devices in the following fields:

(1) Diagnostic machinery—especially devices for locating in early stages deep seated internal malfunctionings, without surgery, and for diagnosing severe foetal malfunctions or defects.

(2) Sceptrology—machines to help people with limb defects (especially leg defects) to lead a more normal life. This includes not only powered artificial legs and stair-climbing wheelchairs but also all kinds of devices for elderly, blind, arthritic, and limbless people to help them perform normal life operations in their homes.

(3) Implant technology—the development of synthetic replacements for parts of or whole internal organs, joints and even perhaps eventually muscles powered by the oxygen and food-fuel in the bloodstream. Synthetic eyes, hearts, kidneys, lungs, liver may eventually be made and means of making the body accept and operate them be developed.

(4) Surgical tool technology—sewing machines can be invented to replace the laborious hand sewing and knotting used at present. We can envisage telechiric micro-hands so that the surgeon feels and sees as though he were operating on a body enlarged tenfold. He can then operate on a patient from a control table outside the sterile operating chamber and use several pairs of hands in turn leaving each pair clamped as he changes to the next. The hands can be stronger than human and yet very much smaller and able to operate beyond normal vision since the television camera can see round the back by a hooked optical fibre bundle attached to the 'hand'.

(5) Hospital and nursing equipment—many inventions can be made to free the nurses and hospital staff from heavy and unpleasant mechanical tasks so that they can concentrate on serving the patient. Hydraulically operated convertible beds, devices for lifting inert patients onto trolleys, turning them over and changing sheets, are already being worked out in my laboratory. Devices for distributing hot food and medical supplies, for hygienic and private bedpans, for the handling and incineration of solid refuse, are all needed.

INVENTIONS FOR MAKING HUMAN WORK MORE INTERESTING

One of the most unsatisfactory consequences of the Industrial Revolution has been the extensive replacement of craft skills in manufacture by subdivision of tasks to the point where one person repeats a few movements every few seconds or minutes. This has largely taken the pride out of work and made most jobs in a factory completely incapable of providing a person with a sense of purpose in life. Now we cannot put the clock back two hundred years and return to handcraft manufacture of all products because:

(1) the number of people in the world is too large for them all to be provided with the necessities of life without the use of power-using machines,

(2) many of the things given to the ordinary individual by the machines are too valuable to be lost (for example travel, comfortable housing, good clothes, hygiene, sanitation and preventive medicine, education and communications, leisure).

Hence, one way out of the dilemma open to the inventor is to finish the job and invent machines to do all the routine work of production, assembly, packaging, testing and movement in the factory. When millions of identical components like electric light bulbs are made or where one is handling liquids and gases in large quantities, as in an oil refinery, the whole factory can be laid out as an automated unit. Where, however, the human operator remains at present the most economical system, the robot with sensory adaptiveness can be developed.

EDUCATION AND COMMUNICATION

These areas are being covered intensively by many large organisations but there is still room for the untrammelled individual inventor who sees the real human needs rather than the immediate marketable possibilities or the simple extrapolation of past trends without considering the limits set by world resources or human physiology.

It is possible that machines can be invented which will relieve the human mind of the burden of remembering facts, rare words in one's language, and the whole dictionary of foreign languages. The problem is to give the human mind such good access to the fact store, encyclopaedia or dictionary that man can use the materials for his own creative thinking, invention, poetry or prose composition, as though he had access to them in his own mind. Perhaps we can invent portable computers which can translate our own language into a universal language, which can in turn be translated by a foreigner's computer to his own language, but no doubt we shall have to speak a simplified basic English to avoid multiple interpretations and will only be able to communicate comparatively simple ideas and facts in this way.

The dictation typewriter which produces a phonetic account of the words dictated into it which can be corrected by the speaker who can then read back directly what he has said, will be a valuable invention. These problems are to gear it to each person's pronunciation and to convert the final statement to correct spelling, although the latter difficulty will not occur with the phonetic languages like German and Russian. Perhaps we can save all newspaper by developing a system where we can read any page of news on our television screens by dialling a code number.

In education it is possible that inventors will finally solve the problems of television cassettes or filmstrips which can give a lecture or lesson so well that teachers can concentrate on the real task of teaching, which is discussion with the pupils and finding out how much they have understood.

TRAVEL AND GOODS TRANSPORT

For goods transport the inventions of the future will have to lie in the field of reducing fuel consumption, noise and pollution, and getting goods transport out of the way of pedestrians and human travellers. Speed is not in general important so we shall see a return to canal use, for example for coal transport, and here the problems are to increase the tonnage throughput of a canal. This will require the invention of better ways of lifting barges up and down hill than the series of locks or lifts used in the past. Similarly, the use of railways for all long distance transport of goods at present conveyed by road vehicles will involve many inventions to improve the handling of goods in containers of all sizes and their distribution by lorry from the nearest station.

For goods transport with fairly low fuel consumption across country where there are no roads, and even for passenger transport, we shall probably see the re-birth of the airship – probably filled with helium. Here inventors will be needed to provide structures flexible enough to withstand differential wind forces, to reduce fuel consumption by using boundary layer propulsion and to control ascent and descent without loss of helium – perhaps by using small separate bags of hydrogen. Inventions to moor the airship safely in special artificial valleys and to cope with the transport of people to and from the airship will also be needed.

Probably we shall invent a way of conveying and distributing food and all regularly used consumer items like soap, newspapers and laundry by a miniature automatic branching conveyor system running in towns in conduits or tubes under the pavement, so that any container can be loaded in at one place and coded to arrive at any other on the system.

It is when we turn to the inventions needed for future developments in public transport that we find the greatest divergence between the genuine best interests and needs of the individual and the pressures of the affluent societies. The affluent society is concerned with providing ever increased speed for the business man – a need which will cease to exist once we have developed the video telephone so that he can hold conferences with people all over the world and see them as he speaks to them. We are already running into troubles because the natural body rhythms and temperature adaptability of the normal human being cannot cope with the changes induced by long aircraft flights, so any further shortening of the air time will increase the strain on the passengers to a level intolerable for all but the most hardy. What the ordinary person needs is opportunities for travel which are safe, reasonable in speed, non-polluting, silent and use so little energy that everyone in the world can get out of their own country at least once in a lifetime. This certainly means many new inventions in the fields of public transport by rail, road, sea and air.

In the case of rail transport, the solution in countries which already have railways might be to use air bearings or electromagnetic suspension, so that we can run trains at speeds up to one hundred miles an hour without wheels, but supported by the existing rails in such a way that they cannot overturn, and with electric traction by linear motor. In the case of buses for short distance transport, the problems to be solved by invention are the reduction of the fuel consumption per passenger mile to a still lower figure, to eliminate pollution by smoke and other products of incomplete combustion, and to reduce noise. In a world with a severe shortage of liquid fuel (which will remain by far the most convenient fuel for road and air transport) we can hope that some new inventions will produce really good improvements in the electric storage battery or in the fuel cell operating on air and a hydrocarbon. In any case the only private vehicles allowed in towns will almost certainly be electric ones, or perhaps a hybrid vehicle with a prime mover engine running at constant speed charging a battery or flywheel.[2]

REFERENCES

1. Intermediate Technology Development Group Publications Ltd., 9 King St., London WC2E 8HN
2. Anon., *Scientific American*, Dec. (1973)

4 THE ART OF THE INVENTOR
Part 1, M. W. Thring

TRAINING ONESELF TO BE ABLE TO INVENT

If we return to the analogy of chapter 1, of a sportsman who aspires to be an Olympic champion, it is clearly necessary to train one's natural inventiveness very rigorously and continually for years, if one wants to achieve the ability to produce really creative solutions to the problems one sets oneself. In this section I shall try to formulate some of the training rules for a would-be inventor.

First he* has to develop in himself the *inventor's eye*. This means he learns to look at every operation or construction around him and think 'why is it done that way or built that way' and whether there is a better way of doing it. This applies to buildings, bridges, tools, kitchen utensils, factory machinery, fireplaces, chimneys, aeroplanes, ships' propellers and to one's own body considered as a mechanism (chemical, mechanical, structural) doing all kinds of jobs with all kinds of tools, and to waterfalls, waves and winds as usable sources of energy. He also thinks about and experiments with every manual job he does, such as washing up, digging in a garden or sawing wood, to see if he can do it better, for example so as to be less tired, so as to use less of his own energy or less water or electricity, so as to balance the use of the muscles on both sides of the body and to do it quicker. This work is for the inventor the exact equivalent of the work in the laboratory for the physicist or chemist.

This development of the inventor's eye not only teaches one to think things back to first principles and to think what are the objectives of the designer, but it also stocks one's brain with all kinds of useful pieces of practical information. When I was working on robots I used to study every job I did with my hands and think whether a computer-brained robot could do it. I gained such an admiration for the trained human hand–eye coordination and adaptability to unexpected circumstances that (for this and other reasons clear from chapter 3) I have now decided to concentrate on telechirics (see chapter 10) where man's skill is available all the time.

*I am using the term 'he' to denote Mr or Ms as women are every bit as inventive as men and often more likely to invent the common-sense everyday things that humanity needs.

The second quality he has to develop in himself is self-confidence to the level of brashness, to the point where he is convinced that he can succeed where everyone else has failed and where all his friends will laugh at him for tackling such an impossible task. Children have this self-confidence and a would-be inventor has to retain it all through his education or, if it has been crushed, he has to restore it. This can best be done by practising inventing small things to improve one's daily activities – gadgets for one's desk, one's kitchen, one's hobbies or for making it easier to use one's tools in one's home, whether it is a flat or a house. By constructing them, one also learns that the path from the first idea to the final successful realisation always involves considerable modification of the first prototype. The examples given in chapter 8 can also be used to develop self-confidence in one's ability to invent.

The third quality that a successful inventor must cultivate in himself is a persistence sufficient to take him past the failures which are an essential part of the successful development of a novel idea. There will always be times when one has failed to foresee some serious difficulty which is revealed by experiment, and the inventor has to see clearly but without giving way to despair whether this difficulty is a basic fault which makes the whole idea impractical, or whether it can be successfully overcome by redesign or a subsidiary invention. There is usually a time and money limit pressing hard to end the experiments and the inventor is the only person who can really see the right decision. He must be very honest with himself: on the one hand not pig-headedly refusing to accept the inevitable; on the other hand not allowing his faith in his idea to be destroyed by his own depression, when the idea is really sound. Being forewarned that such bad moments are inevitable is the best way of being able to cope with them when they come. One can also acquire this persistence by strengthening one's feeling that the thing one is trying to invent is humanly, socially and morally right – by making use of what William James called the 'moral equivalent of war' to give oneself the kind of determination one has in wartime. This is why it is so necessary to choose a problem, such as those discussed in chapter 3, which one is absolutely certain is humanely desirable.

Next the inventor has to learn how to handle his own creative faculty – he has to learn to know himself. First he has to learn how and when he is most creative. All creative people can find periods when they are in a supernormal state, where the brain is most fertile with new ideas and the normal inhibitions, tiredness, laziness etc. are swept away in a flood of energy released by enthusiasm. This occurs at different times of day for different people. Some wake up in the middle of the night and have ideas, others are at their best early in the morning or even around midnight. There is a connection with physical factors, such as a relaxed (but not twisted or slouched) posture and good support in the small of the back, with not having too full a stomach, not being too tired, with plenty of walking exercise in the open air. Walking may even be the best way to achieve the creative state. All these help, but of course do not automatically produce the desired state.

The recent experiments reported in the *Scientific American*, in which it was shown that the practitioners of transcendental meditation did reach a state in which the EEG showed that the brain waves were more regular and relaxed, show clearly that one can learn to produce different mental states by self-control. In a similar way the inventor has to study how to reach his most creative state.

An essential characteristic of the creative state is that one's critical faculty which normally inhibits all new ideas from being formed is completely switched off. Thus ideas are born, many of which will be killed when the critical faculty is switched on again, but others can be followed up by further ideas which meet the criticisms if one is still in the creative state. Thus one has to learn the knack of switching off one's critical faculty at least till some promising idea is actualised by a sketch on the back of an envelope or a few words describing where the solution is to be found. This is done by a conscious effort of will – one says to oneself 'Is there a possibility in that direction?', and refuses to see objections.

There is a legend that Kekulé conceived the idea of the patterns of carbon atoms in the benzene ring when he was slightly drunk; sitting on the top of a bus he had a vision of serpents biting their own tails. There is no doubt that a little alcohol can help to relax inhibitions but for most people it also probably takes the cutting edge from their thought process so that the ideas they have seem marvellous at the time, but do not survive the first serious considerations in the cold light of day.

The inventor has to acquire the art of accumulating the necessary emotional energy to be creative. This means winding oneself up to a pitch where one feels it is as vitally necessary to solve the problem as if one's life depended on it. This is why the really great inventions can only be made by people who can see clearly how necessary they are. Most people find that they can only reach the creative frame of mind when their bodies are really in tip-top condition.

There is no doubt that the inventive frame of mind requires a plentiful supply of the emotional energy which enables one to take a positive outlook on problems and not be depressed by negative aspects (such as the seeming impossibility of the problem, or other people's discouragement). It is true that people in a poor state of health or a condition of over-tiredness or stress have, on occasion, summoned up the necessary emotional energy to achieve an invention or other creative act. Nevertheless it is quite certain that if one wants to increase one's chance of making such an emotional achievement, one must look after one's body as carefully as an athlete preparing for a race. Sleeping pills, excessive alcohol and even stimulants such as strong coffee have a long-term cumulative effect in dulling the keen cutting edge of the thought process. Going for a regular daily walk of a few miles (preferably with a dog, among trees) certainly can help the body to renew the supply of the necessary emotional energy. Too much rich food certainly clogs the mind as well as the body. My own experience is that a heavy meal is definitely inhibiting to the creative state and that when I was on a light diet at a health farm I was more able to have useful ideas and write papers requiring clear constructive thinking than I am normally. This is similar to the experience of many people who have had to force themselves to think clearly about an exceptionally difficult problem.

The fifth quality the inventor must develop in himself is the ability to think every problem back to first principles; he must be able to make a very simple model in his mind of the system he is trying to improve. This model must contain all the essential elements but nothing else. One cannot invent with a mind cluttered with irrelevant details or with a mental model so complicated that it cannot all be grasped simultaneously in one's mental picture. One must know the essential basic scientific principles so that one does not attempt to transgress them (we all receive perpetual motion inventions regularly from hopeful people who have never heard of the laws of thermodynamics) but equally one must not use them as a bludgeon to destroy a half formed new idea.

Finally, it is quite essential for the would-be inventor to develop the ability to 'think with the hands'. Chapter 7 is written to help people to develop this ability which all the great inventors acquired by playing about in home laboratories or workshops. Only the pure mathematician can do creative work without this ability, and the inventor can never invent anything that can be turned into reality unless he has this ability.

THE PRINCIPLES OF INVENTION

I shall now try to enunciate some principles that I believe apply to the act of invention and which the would-be inventor will find useful in his path of progress.

(1) A problem requiring a truly inventive solution looks impossible to everyone except the truly creative inventors, but those problems which can be solved only appear to contradict the laws of science or the known properties of materials because they have been wrongly formulated. Thus the first job of the inventor is to work on the formulation of the problem and see if he can find a formulation which is not impossible to solve but which still satisfies the real aim. See the next section of this chapter.

(2) It is possible to start with a discovery of some natural phenomenon and look for a use for it, but a much more socially valuable method of inventing is to start with a human need and try to invent a better solution to it. It is my definite opinion that unless one has a clear idea of some important human need which may be satisfied by the new knowledge, it is not worth struggling to find one. One example of a discovery without a problem is the silicone 'bouncing putty' which has the exciting property of being perfectly elastic for very fast impulses, but flowing like a very viscous liquid over a time. In spite of competitions and searches for twenty five years, no one has found a worthwhile use for it. Another example is provided by the work of the

Dutchman, Reynst. When he was a student at Zurich he happened to drop a lighted match into a jam jar containing some methylated spirits at the bottom. He noticed that instead of a steady flame, combustion took place with a series of small explosions alternately drawing air in through the mouth and blowing a flame out again. Quite soon he found how to produce sustained oscillations of this type and had the idea that one could get high intensity combustion and high velocity combustion products without an air blower. One could also get very high heat transfer rates to the walls of the combustion chamber as the gas velocity oscillation had a very large amplitude. He spent the rest of his life trying to develop practical applications of this observation, especially for boiler firing and aircraft propulsion. He was not successful, partly because of conservatism in these industries, but probably much more because the advantages of constant volume combustion without a compressor were more than outweighed by the disadvantages of very high noise levels, incomplete combustion and poor control of aerodynamic flow patterns. This sad story illustrates not only that one should start with a very clear idea of an important human problem and not with an interesting observation, but also that in invention it is very dangerous indeed to concentrate solely on one project, since there is always a large element of the law of accident in the development of any invention to the commercial stage. I have known two other inventors who have died disillusioned because their monomania did not succeed.

All the cases where a new scientific discovery (such as the fact that one could cause nuclear fission by neutrons which were also emitted from the fission) or a new observation (such as the formation of glass beads from a fierce seaweed bonfire on sand) has led to a successful new invention were cases where an obvious human need had been until then inadequately satisfied. Nuclear fission by a chain reaction immediately offered the possibility of an almost inexhaustible source of cheap energy (a possibility which is not yet realised thirty years later) while the human desire for jewellery was as great at the time of its first invention, several thousand years ago, as it is now.

Thus this chapter is exclusively concerned with the method of invention which both authors have used, namely to start with a clearly defined human need and proceed to invent a device which solves it in a way that forms a discrete step forward in the direction of an improved satisfaction of that human need, or of satisfying it with less objectionable side effects like pollution, noise or energy over-consumption.

(3) It often happens that there are two or more quite distinct solutions to an open-ended problem: this is one way in which creative inventions are more interesting than logical problems which have only one solution. These two solutions will, in general, have advantages and

disadvantages relative to each other. In the case of the fixed-wing aircraft, the biplane and the monoplane survived together for about three decades before the rising speeds and improved structural materials gave the monoplane the clear advantage. There was similar competition between the radial and axial flow compressors for aircraft gas turbines; now the need for fuel economy has led to the axial compressor with air-fan. Other examples are the Siemens Martin open hearth furnace and the Bessemer converter for steelmaking. Up to about 1950 the open hearth was the preferred instrument for tonnage steel, but now the oxygen-blown converter has swept the board because of the enormous throughput for a single furnace. A final example is the spark ignition and the diesel engine: both have their fields of use, but oil shortage and lead pollution problems will probably cause the diesel to become much more common.

The conclusion for the inventor from this principle is that it is often worth looking at the less successful alternatives to see if he can invent an improvement which will make them more suitable for the changing world needs, or if they suggest a new line for him to explore.

(4) When a major invention is ripe, several people will be working on it at once and may indeed reach it. This occurs because the general human need was clear to many people and the necessary physical phenomena and materials were known although, of course, in many cases the full scientific theory is developed long after (e.g. the invention of using glass). Edison and Swan were both working on the electric filament lamp and achieved practical success independently. Parsons and Rateau were both successful with the steam turbine around the start of this century and, similarly, Whittle and inventors in Germany and Italy reached the successful gas turbine quite independently in World War II. Many inventors have been working on the rotary equivalent of the reciprocating piston for internal combustion engines, but only the Wankel system has had enough work done on it to reach a commercial form. I would conclude that it is not feasible these days for an inventor to work in one of these popular fields unless he has the backing of the kind of money that existing companies spend on their whole research budget. My own work on the telechiric miner will not be successful unless I can obtain a budget of six figures for several years.

(5) It is not possible to budget or programme for the obtaining of a truly inventive idea. There may not be a possible invention or maybe the big research laboratory will only be able to see the trees of the present wood and not the different wood. On the other hand, a big research laboratory may produce quite an unexpected new idea which will open up a field quite different from the one intended. However, when an idea has been produced that looks promising, then the development of it must be very carefully programmed and budgeted

49

and continuously reviewed, as discussed in chapter 9.

(6) The truly creative inventor will necessarily have black moments of frustration and despair, both before and after he has had the primary inventive moment. He must learn to retain his enthusiasm in spite of these.

(7) The principle of *divide and conquer* relates particularly to the run up to the inventive moment. One must learn to pick out the centre of gravity of the problem and continually try to find better ways of doing this. Don't try to solve all the problems at once.

(8) The principle of *successive approximation* applies after the inventive moment. Do not crystallize the idea more rigidly than is absolutely essential to proceed to the next stage. Keep your options as open as possible. This same principle of successive approximation applies to the artist painting a landscape: he must cover the whole canvas roughly with an outline sketch before he puts even approximate colour masses in. One puts detail in as the third approximation. Do not answer the questions in the wrong order; for example, do not bring economics in before the engineering is clear. One must not even think about the next question until one has a definite answer to the one before. The right order for the questions after one has had an inventive idea to solve a definite problem is

(1) Is it contrary to the known laws of science?
(2) Can it work in practice and achieve the desired aim?
(3) Can it work with reasonable speed, rates of throughput, size, number of elements?
(4) Can it be made with known materials and known methods of construction?
(5) Can it be reliable and easily maintained?
(6) Can it be controlled, regulated and adjusted as necessary?

Only then does one begin to consider the economic questions

(7) Can it be made reasonably cheaply?
(8) What are the running costs, labour requirements and maintenance costs?
(9) How long will it last?
(10) How many will break down and how often, and will such breakdowns be disastrous?

The economic questions must all be answered together because they are interlocked; for example, one can make it more expensively to last longer or to have fewer breakdowns.

CHOOSING AND FORMULATING THE PROBLEM

The first step in invention is to decide what one is going to invent. I personally prefer to have a whole group of problems which I have decided are worthy of having inventive new solutions found for them. I have a large notebook with one page for each problem and I make rough notes defining the problem and carry these problems at the back of my mind so that as I look at everything with the 'inventor's eye' I do occasionally get solutions suggested to one or another. At the same time, I also enjoy having a real think about one problem that is interesting me. I discuss various of these problems with other people as often as I can find someone interested, both because merely restating it and answering their questions often causes one to think of a solution oneself while they may make suggestions that show one that there is a whole line of further possibilities. They may even, of course, invent the solution, in which case it is very necessary to give them full credit for their contribution. It is particularly valuable if one discusses the problems with laymen or people skilled in branches of science or engineering other than one's own. This can produce a cross-fertilisation of ideas from a different field, or a suggestion from the methods of earlier civilisation – or even of primitive man – which has been neglected by our industrial revolution, with its wasteful attitude to cheap energy and raw materials and its acceptance of pollution. No inventor can be sure of having all the necessary fields of knowledge in his own brain or of automatically coming across all the devices which can suggest potential solutions.

One may be lucky in encountering a written description of a scientific observation which gives one a clue to a solution of a problem that has been at the back of one's mind. I was not obviously aware of my desire to find a more direct way of converting combustion energy to electricity, but when I discovered an account of the fact that fine silica particles shot from a glass nozzle by compressed air carried a negative charge with them and so constituted an electrostatic generator, I immediately connected this with the possibility of converting the kinetic energy of a jet of combustion gases to direct current, which eventually led me to magnetohydrodynamics (see chapter 10).

My personal experience has always been that the Victorian idea of hugging an idea to oneself and not disclosing it to anyone until it is patented and developed proves utterly unfruitful in the second half of the twentieth century because the danger now is not of someone stealing one's invention and developing it successfully, but of everyone who could help you to develop it being too steeped in existing methods to consider it worth putting any effort into the new idea. For example, I have been preaching the necessity of developing a mole miner (see chapter 10) for fifteen years, but have always so far been unaided by the mining engineers who could not envisage obtaining coal from a deep mine without men on the spot. Even the Russians' ability to control, from the Earth, a robot on the moon to take moon samples has failed to convince mining engineers. The idea will only be developed if the Coal Board or a government ministry gets so desperate for coal that they will try what they regard as an almost hopeless last resort.

One must never allow oneself to be guilty of the well-known 'N.I.H. factor' (Not Invented Here). One must be prepared to co-operate with and acknowledge the work of others who have something valuable to contribute. The object must be to produce a good solution to the problem, not to produce one's own private personal solution. Other people may be forced by their jobs or find their interests concentrated on solving one particular problem, or following one particular field of invention and may prefer to work by other methods than the quiverful of problems method that I use. However, the use of the inventor's eye and the frequent airing of the problem in lectures and discussions is still very valuable. In any case, one has to concentrate on a particular problem before one can collect the necessary energy for the inventive step.

When a problem has been chosen, the second step is to make a first formulation of the rules of the game. These rules are the answers to the following questions: (i) What is one's main objective – is it to use less fuel, fewer materials of construction, to achieve something new like flying for the first time (it is exceedingly difficult to envisage any new inventions in this field that will be of real benefit to humanity), to make something that is much safer, much simpler to construct, will last much longer or require much less attention and maintenance and be much more reliable? (ii) What are the subsidiary objectives, e.g. to be at least as good as the present system in respects other than the main objective, or not to cost more than 20% extra when the main objective is to last three times as long? (iii) What are the restrictions and limitations to the solution? It is in this field that the original statement is most likely to be too restrictive and by seeing that one of these restrictions can be relaxed, one is most likely to find an original inventive solution to the problem. A clear formulation of what one is trying to do takes one half-way towards the invention because, so often, one can reject completely a condition which has become traditional in the existing system and, on other occasions, one can see that one's first thoughts are along a line which is quite unnecessarily restrictive. This is particularly true when the rules of the game have been formulated by someone else, whom we can call the customer. We can formulate this basic principle of invention – *the customer is always wrong*. He always makes the rules of the game more severe than they need be, perhaps because his conservatism leads him to hope subconsciously that the inventor will fail, or perhaps simply because he is incapable of seeing that the accepted rule is unduly severe. Thus invention cannot be made unless one first spots which of the rules can be made more lenient without upsetting the true objective. Often the requirements the customer makes are economic ones – for example, the invention must be made of cheap materials and be suitable for mass production. The inventor must overcome this by the principle of *divide and conquer* (No. 7). In other words, he decides that he will select one part of the problem specified by one rule, as the basic one, and work on this till he invents a good solution to this 'centre of gravity', and not worry about the other rules until he has achieved such a good solution.

Thus the inventor must keep on reformulating the rules until he has picked out the most important problem to be solved by an invention and until he has relaxed all those rules which are unnecessarily severe.

There are examples where inventors have described this stage of the inventive process. Edison gradually formulated the idea that one could use a high-resistance filament heated by a low current to replace the high-current arc lamp which required heavy copper leads from a nearby generator. A Royal Commission in Britain, encouraged by the gas lighting industry, reported that what he was trying to do was contrary to Ohm's law, but Edison's understanding (through his hands) of current networks, as a result of his telegraph experiments, was more accurate than that of the Royal Commission, and he proceeded with the immensely difficult task of finding a filament which became hot enough to emit bright light, without burning out too quickly.

Parsons knew that it would be possible to make a continuously rotating steam engine without valves and pistons by getting the steam to exert a continuous force on blades, like that of a windmill, but he knew that the steam had to expand as it dropped in pressure, so that the inventive problem was to find the right shape of fixed and rotating blades.

THE MOMENT OF INSPIRATION – THE CREATIVE ACT

The third step in invention is to charge oneself with sufficient emotional power to be able to switch off one's inhibitive critical faculty and produce a new idea. One has to think about the human value of solving the problem and answer fully all the doubts one has about it, not by suppressing them, nor by ignoring them, but by acquiring confidence that one can find a solution which will not be in any way subject to these doubts. The analogy given in chapter 1 of the rider and horse in a field with high hedges all around applies particularly to this stage: the leap over the hedge can only be made when the rider and the horse both feel quite convinced that they will get over this time.

Having achieved the necessary force of determination, one has to empty one's mind, as far as possible, of preconceived notions, knowledge of how the problem is solved at present, and the obvious small improvements. I was told by Dr M. Pirani (who invented the Pirani vacuum gauge early in this century and had dozens of other inventions to his credit in the field of electric heating and lighting and refractories) that he never read the literature when he undertook a new problem; he only read it after he had worked out a new idea of his own. Barnes Wallis has said the same thing. It is certainly not the easiest way to be creative, to search all the relevant literature and see everything that other people have tried. If one does this, one is liable to be completely put off by the conclusion that all possible ideas have been tried. On the other hand, when one has arrived at a worthwhile idea, one often finds that several other people have tried the same idea. In this case, one says to oneself 'That proves the idea is a good one, and I will be the person who overcomes the obstacles that have prevented others from pushing this idea to success.' If one has developed sufficient mental self-discipline it is possible to skim through the literature and read of other people's attempts to solve the problem without being inhibited or put off by their failures. It is essential in this case however to make strenuous efforts to look widely and not get caught up in details and to retain an open mind to look for ideas from quite other fields.

Similarly, one must erect barriers against all the outside people who tell one it is insoluble, that all the ideas worth having have already been had, that one is not clever enough to solve this problem, that one is tackling too many problems at once.

Now we come to the actual inventive moment.

Roughly speaking, one can distinguish four different methods used by various inventors to achieve the moment of inspiration.

The first can be described as bullying oneself – setting a time limit and forcing oneself to maintain close attention on the problem until an idea of a possible solution presents itself. This was Edison's method with the lamp filament. He used to lock himself away in a tiny cupboard for many hours. It requires an intense mental effort; every time one's mind wanders away, because a difficulty blocks the line of thought, one has to force it back. This method requires the development of an exceptional degree of mental self-control and may be too difficult for many people who are nevertheless capable of really original inventions using the other methods.

In practice the engineer may often find himself in situations where he has to produce an inventive solution to an urgent problem. In this case he has to make extra efforts to switch off his inhibiting internal faculty so that he can produce any solution that works rather than struggling to find the best one.

The second method may be called the method of gestation. This means that one has a formulated problem for which one cannot see a good solution but for which one feels sure there is a good solution. One keeps notes of it preferably on a single sheet and reformulates the rules of the game from time to time as a result of pondering about it or discussing it with others. This gestation process may take weeks or months until eventually 'the penny drops' or the light dawns and one gets an idea that makes the whole problem soluble. This method of gestation is not, of course, suitable for solving a problem on which there is an urgent time limit, since it may take many months or one may never hit on a feasible solution. The matter of finding a successful inventive solution to a problem is always subject to chance and all one can do by using any of these methods is to increase the probability of success from a vanishingly small value to a reasonable value, perhaps even as high as one in two. However, if one wants to have a reasonable chance of producing a worthwhile inventive idea within a few months, one must supplement this method by one or more of the others.

The technique of serendipity can also be very useful in this case. In its most extreme example one goes to a library, looks up the book closest to the subject of the problem, and then goes to the appropriate shelf of the library and glances at random through books on the shelf other than the one concerned.

The third method of finding an inventive solution to a difficult and important problem has been widely publicised. It is the method of synectics or brainstorming. A group of people sit together in a room and toss the problem around using free association and freedom from critical inhibitions, both of which are essential to original thinking. Again, the more widely spread their professional training and non-professional interests the better. They must reach a co-operative atmosphere where one person's casual idea is seized upon and developed with further ideas by others, to give it every chance of proving a feasible solution to the problem. This method of invention is more suited to extroverts, whereas all the other methods are favoured by introverts. The resulting invention should really be attributed to the whole group. The method is certainly rapid since, if nothing is produced in two or three sessions, it is unlikely that the method will be fruitful.

The fourth method is the systematic or long shelf method. This means essentially making some kind of table or list of all the possibilities and then thinking about them one by one. If any possibility looks promising for the solution of one of the preselected problems, it is then carried through the preliminary development stages described in the next section of this chapter.

Another variant of this method is to carry out all kinds of experiments in the laboratory without a clear idea of where the solution to the problem lies, but in the hope that an observation will give one the clue to the solution. In some cases I find it useful to go into my workshop and make a mock-up model from easily worked materials, working it out as I go along.

There are also three hints which one can give the would-be inventor to help to achieve the moment of inspiration. The first is to learn how to switch off one's critical faculty so that half-formulated ideas are not strangled by destructive criticism before they can be fully formulated and converted into schemes sufficiently defined to enable the difficulties to be solved. This means being entirely positive and receptive to new ideas whether they are one's own or are suggested by a remark of someone else; it is associated with the optimism that believes there is a good undiscovered solution to the problem and that it can only be achieved by struggling against the pessimism and hopelessness which lurks like a black shadow in everyone's emotional make-up. One must never allow oneself the luxury of giving way to despair, since this quickly destroys all one's creative power as a breach in a dam allows the water to flow away.

The second hint is similarly concerned with the building up of creative power. When one is really determined to solve a problem, one will have many ideas which seem at first sight to be the perfect answer but which prove (when taken through the stages described in the next section of this chapter or even when developed as described in chapter 7) to possess insuperable difficulties or disadvantages which make them less effective than existing solutions to the problem. Such discouragements have occurred to all inventors and the successful ones are those who have raised themselves up again and again after their abandonment of pet ideas until they eventually found the solution they really wanted.

The third hint is well expressed in the phrase 'divide and conquer'. Do not try to produce a complete solution to the whole problem all at once. Pick out the 'centre of gravity' of the problem – the key difficulty – and concentrate all your efforts on that, ignoring all the other aspects. If you can produce a real step forward in the key difficulty you will usually find that, at first, it seems to break other rules of requirements but that you can make subsidiary inventions to clear up these difficulties. A good example is the fact that the combustion gases are a very poor electrical conductor and this gives the magnetohydrodynamic generator a very high internal resistance: then one thinks of seeding the gases with readily ionised alkali metals. To do this, one must look carefully at all the objectives and see which is the most important (for example, to reduce the fuel consumption) and concentrate entirely on this one.

CRYSTALLISING THE IDEA

When the inventor has switched his critical faculty right off, and has caught a glimpse of a possible solution to the problem, he must proceed very carefully to avoid, on the one hand, losing sight of the solution and, on the other, crystallising the idea so precisely that it develops in only one of the many ways possible.

The basic principle by which the inventor must proceed is that which is called in mathematics 'the principle of successive approximation'. The first thing he must do with the idea is to try to formulate the essence of it, either by drawing a sketch on the back of an envelope or by writing down a few words as a reminder of the field of knowledge where a solution might lie. It is very important to cover the idea as broadly as possible so as not to exclude any area, and to include no detail at all, so that he feels no compunction in throwing away half a dozen formulations or sketches, since he has put no careful working out into them. It is very much better to use rough pencilled notes than a ruler and compasses. The same principle applies in all creative work; for example, in painting a picture, the artist must sketch roughly all over the canvas at least once, and preferably twice, before going into any detail at all. So with a new idea one must look at it all over and from all sides with half closed eyes before one begins any kind of detailed work.

At this stage you must 'keep all your options open' because when you have made a feasible rough sketch you must switch on your critical faculty. You do not yet concern yourself with any questions of economics, such as reliability, maintenance, labour, cost of the materials of construction – all these come much later, but you do ask 'can it work, can it be made with existing materials, can it give a better solution to the key problem?' You ask the primary engineering questions and only if you can formulate the solution in such a way that suggests you have answered these do you proceed to the second stage —drawing-board, and the calculator or computer —and begin to add the details of dimension, speed, lubrication, method, shape and optimum materials, and work out static and dynamic forces and stabilities, efficiencies and other key factors. Even these calculations should only be done so as to give the answers to one or at most two significant figures. One has merely to get the decimal point in the right place at this stage. Again at this second stage, one must be very careful not to go into too much detail, for fear of wasting so much time exploring what proves to be a blind alley that one has no energy left to explore the proper avenue.

If it is necessary to construct models to explore three-dimensional relationships or movements more fully than can be done on the drawing-board, these models must be made of the softest, most easily worked materials, that have just enough strength to serve their purpose. Bent wire, Plasticine, wood, aluminium sheet, structural kits held together with nuts and bolts are the appropriate materials because they can be scrapped or altered with very little effort and so do not commit the inventor to a preconceived notion. It is of course essential that the inventor himself makes these first drawings and models because only he can develop the idea as he does so.

It is also an important application of the 'divide and conquer' principle that one does not at this stage make one component serve two purposes, but always designs each part for one function only.

Part 2, E. R. Laithwaite

KNOWLEDGE AND WISDOM

Of all the living creatures on earth, Man is the only one to make fire, to wear clothes, to have imagination, to invent, to use the wheel – and, it is said, to know that death is inevitable (although the last could be doubtful). There is one other ability (or disability?) of mankind that is unique. Only men use pure metals. For centuries we have convinced ourselves that the wheel was perhaps the one thing that we make that goes one better than 'the Creator', the Designer of all living things. We have persuaded ourselves that we have free will, that we can make what we like and use it for good or evil and we have long personified good and evil and seen them as giants locked in perpetual conflict.

In Science there lies Knowledge, perhaps *all* Knowledge (if only we had access to it), but there seems to be very little Wisdom. We must look for that in the arts, in the ancient civilisations who might even have had more of that rare commodity than the society of the twentieth century. The ancient Chinese monad (figure 4.1) expresses two profound ideas. First it has inherent asymmetry – the concept of left and right-handedness that still baffles the physicists of the 1970s as they argue about particles and anti-particles and about parity – whether the Universe is left-handed as a whole. Second, the dots on the monad were intended as a constant reminder that there is no evil thing which contains *no* good, that every good thing contains a little evil. There is no beauty devoid of ugliness nor anything that is wholly ugly. In the past I have expressed the same sentiment in public lectures, by saying that the only law of the Universe I have been convinced cannot be broken is that there is no pure black, no pure white, but only shades of grey, and that the laws of physics are man-made and all are subject to need for subsequent adjustment.

Are we really *that* clever for having invented the wheel and for using pure metals? Of the latter my colleague Gilbert Walton has written:* 'No doubt if we knew how to do so, we would abandon metals in favour of organic materials which have a far greater variety of properties.'

*Walton, Gilbert, 'Facts and artefacts', *The Modern Churchman*, July 1964, Vol. VII, pp. 233–238.

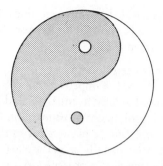

Figure 4.1   The Chinese monad – symbol of good and evil

We have come rapidly, in this century, to the age of plastics and have soon been alarmed by their indestructibility: that plastic food container tossed casually to the ground after a picnic will remain unchanged for a thousand years. Already inventors have made plastics that are affected by ultra-violet radiation so that they finally disintegrate into simpler substances. They have a 'lifespan' as we do. We may be discovering why living things are born to die and that only by so doing is there the possibility of eternity. The history of life on earth extends backwards for hundreds of times the history of man.

It is possible that we could profit by searching *backwards* rather than forwards for all those things we have failed to notice that have evolved rather than been invented, for we may discover that nature study is a 'mirror' to our man-made machines. Where the image is identical we may find Nature's version superior to our own. Where there is only a superficial similarity in the reflection, we might learn how to perfect our embryonic inventions; where there are images in the mirror that do not appear in our world we shall be most severely taxed, for we shall not even know their purpose.

THE HISTORY OF MAN-MADE THINGS

We may begin by asking ourselves such simple questions as: must a house be made of bricks? – to which the answer is obvious. But to answer: must windows be made of glass? – is much more difficult. Transparent plastic scratches too easily. Pure quartz is far too expensive. Yet we can imagine quite easily a substance like glass which has the strength of steel. It is simply a question of the time at which such things are invented. Some of the questions below could have been asked in the past. All seemed at the time, to be answerable simply by 'Yes', but all have since been changed.

Must ships be made of wood or other lighter than water materials? – First iron ship: 1843.

Must shoes be made of leather? – First plastics in the 1940–1950 decade.

Must tyres be made of rubber? – Again a twentieth century replacement.

Must high-power electrical transmission lines be made of copper? –
Cheaper aluminium was thought to be too weak in tension until a
steel core was inserted in 1909.

It is so obvious once it has been done.

Thinking of materials in need of improvement is a good exercise for inventors for it stretches the mind at least as much as finding the solution to many problems. I confess that it took me some time to think of glass as an example, for even glass has been replaced by plastics for optical instruments and the like. Car windows are now usually a glass/plastic/glass sandwich. Only house windows, perhaps, remain to present any real challenge, even in this example.

To ask why an object has remained the size or shape that it has can usually be answered by some such phrase as 'tradition' which is a 'blanket' word for all kinds of complexities, such as the production of the cheapest article, the fact that the cost of re-tooling is far greater than the saving in production costs for as far ahead as the accountant can see. 'Tradition' includes such almost unbelievable mental blocks as superstition, a feeling of insecurity ('I'll *never* get accustomed to this decimal currency'), a preference to do nothing, for which there are no kinder words than *laziness, arrogance,* or at best *complacency* and self-contentment. And in most of these blocks we may find at least an element of that most deadly practice we call 'brainwashing'.

THE FIGHT AGAINST BRAINWASHING

The fact that the word itself was coined with the coming of the totalitarian states and the ruthless methods employed to make people conform is not to be taken as indicative of the fact that it is a twentieth century creation that we fight in order to become inventors. Very terrible tortures have been carried out in the name of religion to say nothing of the verbally persuasive methods, as far back as religion can be traced. What is worse is that there is no brainwashing half so deadly as self-brainwashing. If you invent a new problem-solving technique, the immediate reaction is to try to apply it to everything in sight rather than leave this task to lesser mortals, in order to *empty* the brain and start on a quite new line of thought.

I know of only one positive weapon with which to fight self-brainwashing, although a daily 'mind-cleaning' exercise can become as habitual and beneficial as daily physical washing. The method I have in the past described as the 'cussed approach' involves the mental challenging of almost every statement that one hears or reads. This will lead one to such thoughts as 'Why will iron ships sink? – Let me consider, whether they could be made so as *not* to sink.' It may have been the stimulus behind the famous apple that Isaac Newton saw falling. 'Does the apple fall? – No, perhaps the Earth comes up to hit it!' – Unlikely, but as we now know, it contains an element of that better knowledge that the affair is mutual. Try rejecting Einstein's premise that the velocity of light is constant on all frames of reference and assume instead that it was reducing with time (a vast time scale, such as a 1 per cent reduction in velocity in 10 million years, will do as a start). This will 'build in' the concept of an apparently expanding Universe, but what are the other repercussions of this assumption and do they throw up any new ideas?

Just to illustrate what *can* be done, I believe that in pre-television, pre-transistor days, an engineer responded to a challenge to build a radio receiver whose only power supply was obtained by connecting it to a gas tap! I will not tell you how this can be done, but leave it as a problem on which readers may elect to dwell in odd moments. Unprofitable as a commercial receiver it certainly was, but very profitable for the man who sharpened his brain on the problems he met in completing it.

FASHION, EVOLUTION AND THE MANAGING DIRECTOR

Word association tests would doubtless reveal that most people's reaction to the word 'fashion' is 'clothes', but a moment's reflection is sufficient to suggest other topics that have fashionable aspects, and the extent of these topics is beyond what most of us consciously expect. Furniture and the decoration of our homes are among the more obvious. The arrangement of the average living room underwent an enormous change when the fireplace was made redundant as the focal point in favour of the invention of Logie Baird – the TV. If he had not invented this device I know at least one other man who would have done so very shortly afterwards. This is a facet of invention that can be the subject of a thousand debates. Does a major invention occur when the state of the world's technology demands it? Was the man made for the time or the time for the man?

Builders of industrial empires have expressed to me their amazement at the apparently clairvoyant abilities of the organisers of Marks and Spencer Ltd, who as one business expert put it 'seem to know how to have the articles on the shop counters on the very day the public want them'. Let all would-be inventors take note. Your inventions must be *timely* or you will either be beaten to it at the last moment, or will suffer the fate of Charles Babbage, whose Computing Engine of 1838 was waiting for the inventions of the triode valve and then the transistor before it could dominate the world, and he never knew it. It is a hard road for the man so far ahead of his time.

There is another enemy of the invention ahead of its time which is the creation of the 'self-made man' who rises to power and sets himself up as God. I know a well-known industrial organisation (not British) that was manufacturing the best permanent magnet steel known at that time. A director of that company made a point of telling his visitors with authority that: 'We are now making the best permanent magnets that will ever be made.' The firm's magnet department survived in this case, because it had a research section whose personnel were never satisfied – the organisation was big enough to swamp the individual. Had it not been so, this factory might now be making biscuits!

What of such earthy topics as engineering? Surely we change the form of our machines as new materials, new inventions make possible 'better' designs? It all depends on how one interprets the word 'better'. In 1920 a better electric motor was one with a higher efficiency. Few engineers asked what it cost to ask for 85 per cent instead of 82 per cent. Efficiency was the fashionable commodity of that time. In the rapid advance in technological hardware that followed, power-to-weight ratio (horsepower per lb, or kilowatts per kilogram) was to replace efficiency and in its turn be supplanted by first cost as the 'Age of the Accountant' overtook us. This last is but a personal view, probably not shared by the majority of readers and certainly not by the accountants. But it is my impression that in heavy industry at least, the depth of forward-looking planning extends to no more than a year or two, for the monetary system of today demands that the company's books be seen to balance 'at the end of this year' – or at best at the end of the next.

The managing director has the ultimate responsibility for policy and if he would make his company profitable, he must be as ruthless as evolution. He must kill off all those things which are unprofitable. Unfortunately he usually fails even to approach the effectiveness of nature, for he is never so ruthless, never so willing to admit a mistake and succumbs for one reason or another to the temptations of tradition.

So what have inventors to learn about the dangers from fashion and the managing director and about the laws of evolution that are so terrifying? They must pander to the accountant while the 'Age of the Accountant' lasts. If they would seek to arrest pollution they must invent recycling processes that can be seen to make more profit for the manufacturer than would be the case if he started with raw material. It is useless to organise the collection of newspapers, aluminium milk bottle tops or ensure the return of glass bottles themselves, if it then costs the manufacturer more to make new paper, bottle tops or bottles than it would if he simply cut down trees or mined metal. He will go on cutting down trees until there are none left unless you can make him new machinery that reduces the cost of recycling below that of starting from scratch. The accountant, of course, will claim that it is not *his* fault. So long as a housewife will not boycott a firm that discharges its waste into the river in favour of one who does not, how can he ensure the survival of his firm? The customer will claim that it is not his or her fault – they have a family to raise on a limited amount of money. I am very much afraid that it is the fault of us all that we are not better inventors. There is an inventive streak in all of us to a greater or lesser extent, but it must be developed to be useful and it is to this end that we write this book.

THE CONVENIENCE OF MANKIND

We are slow to accept inventions that are particularly aimed at increasing the *convenient*. At times we deliberately go out of our way to make them less convenient, provided the result is a greater degree of conformity to a strict classification. The most glaring examples of today are first, the replacement of letters by numbers for the first three digits of all London telephone numbers. Kensington no longer begins with KEN, but 589, which does not even conform to the same hole positions on the dial. Second, we are now blessed with Postal Codes, and who has space in his memory for this lot? Third, we have metrication, no bad thing in itself, unless taken to extremes; by this I mean that I have no objection to working in centimetres instead of inches for the measurement and contemplation of common objects such as nuts and bolts and screws but there is a powerful lobby, among physicists in particular, to restrict units of length to (metres)^{3n} where n is an integer. In other words the only possible units to be used in the description of objects are nanometres (10^{-9} m), micrometres (10^{-6} m), millimetres (10^{-3} m), kilometres (10^3 m), megametres (10^6 m), and so on. We must reject this vigorously, for none of these units is *convenient* for the description of a large proportion of everyday objects and it is not a question of familiarity alone that makes 7 inches more convenient than either $1 \cdot 778 \times 10^2$ mm or $1 \cdot 778 \times 10^{-1}$ m. When we replace the duodecimal system (as in pennies in a shilling, inches in a foot) by the decimal system we make measurement conform with mathematics and the argument is powerful, even though the number 12 is a much more useful base than 10, being divisible by 2, 3, 4 and 6. Metrication bars us from that other most useful ability in engineering – to keep dividing or multiplying by 2 as in $\frac{1}{2}, \frac{1}{4}, \frac{1}{8}, \frac{1}{16}, \frac{1}{32}, \frac{1}{64}$, etc.

It is not in the *usage* of decimal measurement that we shall concede something but in our mental *concept* of distance and space. 'Convenient' numbers are those which begin with a single digit in the units column, such as 5·14, 1·37, etc., with no 'ten to the something' tacked on. This desirable feature has manifested itself throughout history by pioneers in separate subjects isolated from other technologists (and of course there was technology long before there was science). Thus the first measurers used cubits, measured horses' heights in 'hands', used the top joint of a man's thumb as an inch, his foot as a foot. There was the desire to make each unit easily reproducible. To measure cloth by putting the hem between the teeth and stretching out one hand to its fullest extent and holding the same hem thus at arm's length gave us the yard. Doing this with rope with a double arm stretch gave us the fathom, for this was the way the early seamen hauled in the lump of lead on a rope when measuring water depth (it is not an accident that 1 fathom = 2 yards). If we lose the inch it is no tragedy. If we then lose the centimetre we shall lose just a little of what I call the 'common touch'.

It is also interesting to note that whilst the whole of Europe is now decimal in calculations, measurements of length and mass and in currency, time remains far from decimal; the more useful number 12, and its half or double, appears in the relationships of seconds to minutes, minutes to hours and hours to days. When shall we have a decimal clock? Likewise we measure angles by subdividing a circle into 360 degrees (a 'hybrid' of 10 and 12!) and then each degree is sub-divided to 'minutes' and 'seconds' as with time. I have never actually seen an angular velocity stated as 'minutes per minute' or 'seconds per second' but it would be amusing to do so in a scientific paper

63

to highlight the inconsistency of modern man who strives for clarity and delights in coining new words.

Even in the very subject of invention itself it is interesting to note that British Patents began with number One in the year 1617 when it was granted to Avon Rathbone and Roger Burges on the subject of engraving and printing of maps, plans and the like. On 26 July 1853 what was thereafter called the First Series was terminated at number 14359 and the Second Series began, believe it or not, with a number One, dated 1 October 1852. At the end of 1852, i.e. 31 December, they began again at One, dated 1 January 1853, there having been no fewer than 1211 patents between September 1852 and the end of that year. Thereafter there was a new beginning at no. 1 every first of January until the end of 1875, all of which were said to be in the Second Series. The numbers of patents per year after the end of 1852 were successively 3045, 2764, 2958, 3016, 3200, 3007, 3000 (they must have celebrated the 'tidyness' of 1859 in champagne!) and a steady three thousand-odd, per year, thereafter. Then began the Third Series, beginning of course at no. 1 again, but going on through the years up to the present time. There are thus no fewer than 25 of each patent numbered 1, 2, 3, ... 1211, and 24 of each up to 2764. In referring to older patents, you need more than just a number!

The first series was recorded in the Great Seal Patent Office. Under the Patent Law Amendment Act 1852 the idea of the Provisional Specification was introduced and the office of the Commissioners of Patents and Inventions was set up. From 1617 to 1858 drawings were filed separately in huge volumes of full-sized draughtsman's paper. The inconvenience this causes us today is considerable. From 1858 to 1876 drawings were included with the specification but there was no restriction on size. From 1876 drawings of standard size only were permitted. How long, we might ask, before all drawings (and specifications, too) will only be acceptable on standard microfilm?*

This section has been not so much a guide on how to invent as a warning of the 'minefield' that confronts every inventor whose products fail to conform, fail to promise financial gain or even revolt against a 'sacred heritage'. He or she may be honoured perhaps a hundred years after their death, which is small consolation. Yet history is filled with accounts of those who have pressed on with their ideas literally 'to the bitter end', so we may take heart. There *is* a satisfaction for those who walk alone, even, it should be noted, if they are destined to be seen to be wrong by the next fifty generations of man!

ASSOCIATION OF IDEAS

This phenomenon can easily be detected either in oneself or in another person, for it gives rise to the almost instant thought (in oneself) or verbal expression (in someone else) which begins: 'That reminds me of ....'

Whenever that occurs it is a potential 'seed' for an invention and all such observations should be followed up mentally to see how far the comparison

*The authors are indebted to Mr Nigel Greatorex for collecting the above information on patent classification.

64

can be taken before it no longer applies. Having found the 'breakdown point' ask yourself *why* it broke down where it did, and whether this was due to some fundamental reason, or simply because the counterpart of one system has never been tried. If the latter be the case, there is the possibility of trying it and the result may be trivial, tiresome or meaningless. If however it should still be interesting, you *could* have your invention.

CO-ORDINATION

From the moment a baby is born (perhaps even earlier) it practises association of ideas. A moment's consideration is enough to show us that this is a necessary survival mechanism that is therefore instinctive. See the baby reach for things that it cannot reach. It is learning how to co-ordinate the images seen by its eye with the length of its arms. Later, it will experiment with how far it has to crawl to reach what it sees. The fact that we never finally complete our education in this subject is demonstrated when it sometimes takes us a few seconds to decide whether a tiny spider that dangles in front of our eyes when we are sitting in the garden is a bigger spider at a greater distance.

The small baby is expected to learn quickly, so it needs a lot of practice in association. The teething ring is not initially put into the mouth because it is known to help the teeth through the gum, nor even because it feels nice. It is to see what a thing that looks like 'that', feels like. After lots of practice a human knows what an object that looks like a sphere 'feels like'. But it is not instinctive. It has to be *learned*.

But as in most other learning processes there are always pitfalls. Many people who have never handled snakes think of them as being 'cold'. They associate their shape with that of worms that *are* cold. Red berries 'look' tasty, green plums 'look' sour, but there are obvious exceptions to each concept. Without co-ordination there could be no conjurors or illusionists. In co-ordination and association there are the beginnings of curiosity – and curiosity is often the beginning of invention.

Association of ideas stretches beyond the conscious. A piece of music being *heard*, a picture being *seen*, can flash into the other senses a vivid recollection of a time past.

HABIT

There is virtually no dividing line between the association of ideas and the peculiar phenomenon we call habit. Habit is a scientific word, having precisely the same meaning as the same word used in its everyday sense with reference to such practices as smoking, drinking, drug-taking, and so on. (This is a rare occurrence incidentally, for one has only to think of the everyday meaning of a few engineering words such as damping, crank, excited, dog-clutch and running fit, to realise how ridiculous scientific literature would appear if read in the context of everyday speech. It is a useful brain exercise to try to think of examples other than the ones above and fit them into sentences that make them most comical.)

The strange thing about the alliance between habit and association is that for the inventor the first is an enemy that must be fought daily, the latter is a friend. You bolt your front door each evening as a matter of habit. Proof of this occurs when you are over-tired, for by the time you reach the bedroom you cannot recall consciously having done it on that particular evening, so you go back and look, sometimes more than once! Each time you find it locked. Actually, if you thought about it on other nights when you were not exhausted you would fail to remember doing it also; it all depends on whether you were concentrating keenly on something else as you bolted it. Habit is apparently not seriously impaired by fatigue. The faculty that seems to 'go to sleep' is a kind of 'auditor' that normally assures us that what *should* have been done, *has* been done, even though unconsciously. Be that as it may, over-tiredness in my own case certainly brings habit to heel and causes me to take nothing for granted.

FEAR

If we know that when our habit faculties (if they can be so called) are inhibited, we are more likely to invent because we take nothing for granted – even become afraid of those things that would normally not cause us to fear them, and yet be capable of logical thought – then we can certainly practise the art of taking nothing for granted when we are less fatigued and in a healthier frame of mind.

One of the very necessary ingredients of an inventive person is the ability to use imagination. Yet imagination is also a powerful ingredient in the experience we know as fear. Throughout history fearful men have been branded 'cowards', and relegated to the lowest level of manhood. But history, as was customarily taught in the schools of forty years ago, was mostly a list of battles and wars. Life was simple to the historians prior to the nineteenth century. There were 'baddies' and 'goodies', right and wrong, friend and enemy. Then we saw 'through a glass darkly, but now face to face . . . .* We know that bravery of the old-fashioned kind can be the result of:

(1) Intoxication
(2) Extreme rage
(3) Intense pain inflicted by the enemy
(4) Drug administration
(5) A state of hypnosis

in addition to the more laudable causes of love, duty, and self-sacrifice in the face of necessity.

One might deduce that no brave man could invent. This would be an

*St. Paul's Epistle to the Corinthians.

inversion of a truth that will certainly not stand up, for one of the most ingenious men I know is also as near to a fearless man as any I have known. I would therefore only seek to encourage those who have inwardly branded themselves as cowards, for I would place a lot of weight on the theory that a person who knows fear has imagination. Channelled correctly by a healthy person this can lead to great inventions and to great rewards both mentally and physically. But the price is high for the imaginative mind will suffer indescribable horrors when the body is sick. There is no logic in fear.

The best of mathematicians can tell themselves that their hallucinations during an attack of a virile strain of influenza do not exist, but they will fear just the same. For a person afraid of moths it is quite useless to demonstrate the harmlessness of these creatures. Fear is allied to habit and is the very bedrock of superstition. The inventor must struggle to rechannel fear into more tangible things.

ASSESS YOUR OWN POTENTIAL

It is difficult to generalise on what kind of background, social, educational, inherited, etc. makes the springboard (if any) for a good inventor. At one time I had a theory that a lonely childhood was a firm contributor, for I can recall the many occasions when I had to play games intended as a contest between two or more people (but especially two) by inventing methods of playing against myself that allowed all the personality touches of two *different* people (sometimes one of whom cheated!). Lewis Carroll makes reference to this in *Alice's Adventures in Wonderland*, including the cheating; on such occasions Alice punished herself for cheating. Whilst I found that my theory that the cream of the world's inventors were 'only children' or outcasts in other ways did not hold up against the many examples of inventors from large families, I am sure that my own loneliness as a child had its effect.

5 PHYSICAL THINKING

E. R. Laithwaite

The standard pattern of science education in Britain in the 1970s is that school pupils up to the age of 15 or 16 have a selection from (or all of) physics, mathematics, chemistry and biology thrown at them whether or not they are interested. Likewise those who are fortunate enough to know their own minds at that age, and would have opted for science, have an equally formidable array of subjects thrown at them (e.g. English literature and language, history, Latin, etc.). In 1968 the Duke of Edinburgh formed the Schools Science and Technology Committee (SSTC) with the aim of bringing school science teaching up to date in a modern technological society. As I myself had expressed anxiety over this in public I was both honoured and delighted to join the Committee, which met once a month for several years, often under the personal chairmanship of His Royal Highness.

In the course of this work I undertook other activities which were relevant to the subject. For example I was appointed chairman of the advisory committee for physics of the Associated Examining Board (then the third largest of nine Examining Boards which set papers at 'O' level and 'A' level). I held this post for six years and during that time I discovered a number of facts which were as new to the rest of the members of the SSTC as they were to me. For example, each Examining Board is a privately owned company and not subject to the policies of the Schools Council. Like any other well-run business, they fear one thing above all else – losing customers! Schools pay handsomely for the services of an examining board, but the easiest way to lose customers is to change the syllabus. Add but one new topic, such as *magnetic circuits* or delete the *tangent galvanometer* and 59 school physics teachers threaten to 'go to another board'. The reason is not difficult to find. Why should a school science teacher have to keep abreast of the latest developments in his subject as well as mark all the homework on a salary which ranks among the lowest in the land?

Another reason why the physics syllabus had remained static was the interplay between examining board, science teacher and book publisher. I could not (I was told at the board) introduce a topic into the physics syllabus unless it was covered by the textbook. A number of publishers were then invited to meet the SSTC and were asked whether they would publish a physics book containing new material. 'Of course not,' we were told, 'It wouldn't sell. There is no syllabus demanding it.' The chicken and egg problem once again!

So what is this stereotyped physics which was taught in the 1960s and by whom was it taught? It was the same material on which *I* was reared with some electronics and atomic physics tacked on to the end to give it an air of respectability and 'with-it-ness'. It was taught, in the vast majority of cases, by teachers who had only *known* physics. At best they had first degrees in physics, at worst they fitted the description: 'Those who can, do. Those who can't, teach!' In a very few cases there were ex-engineer teachers who found themselves tied wrist and ankle by the *Syllabus*.

I would like to believe that we did a good job within SSTC in bringing together under one roof, at least once a year, a Standing Conference of representatives from all organisations who are (or *should* be) interested in school science. In how many other places, after all, do the Trade Unions sit at the same table as the Royal Society? (in their own right, that is).

My purpose in beginning this chapter with this bit of recent history is to highlight several aspects of the background of the average man or woman of today, who could be carrying immense potential as an inventor. A school pupil of thirteen years old will accept almost any dogma if it is launched at him with sufficient force. So we all believe matter to be made up of molecules, which in turn are made up of atoms, and we all know what an atom of hydrogen looks like – an orange in London and a pinhead in Guilford, and the next atom is in North Africa! Fine, but what *shape* is this electron? The answer must be that it *has* no shape, or the human mind at least would very rapidly imagine it cut in half, which we know cannot be. So if it has no shape, how can it spin? In particular, about what axis does it spin? One is forced to agree with the mathematician who is reputed to have said: 'An electron is only properly described by a set of equations.'

So why are we not honest with our school children and tell them of the most fundamental tools which they will use (whether they become scientists or not) for the rest of their lives? I refer to such topics as symmetry and analogy.

ANALOGY

Analogy can be referred to in a somewhat derisory manner as 'another name for a fairy story' in that it is untrue, but its morals are not in question! Whether we like it or not, it cannot be avoided in our daily lives and, when coupled with that all-embracing 'uniformity of nature', which is perhaps the *only* law of physics, it provides the complete guide to life as we know it.*

The stumbling-block without a doubt lies in the *teaching*. Almost by definition (at least *so far* in our development) teaching consists of the handing

---

*The uniformity of nature is the simple concept that those things which have always occurred in a given situation have a very high probability of continuing to occur in any future identical situation. So familiar do some of these occurrences become that our brains relegate them to a separate storage compartment which needs no conscious thought in order to absorb them.

on of facts and ideas. Rarely is it put to school children that 'facts' are generally far from rigid or that an idea, or more specifically a way of thinking about something, which is extremely useful to one individual can be confusing to another.

Of course this makes the setting and marking of exam papers very difficult, especially in quantity. Only at university postgraduate-degree level can we indulge in the luxury of marking each script as if its originator were a human being with sparks of brilliance here, great empty voids there. We are undoubtedly mass producing the 'freshmen' of the universities.

The use of analogy in inventing can be summarised under four headings:

(a) An analogy is a fairy story – a parable, an attempt to make the invisible visible; to make the intangible tangible; in the extreme case to give the fallible human mind the illusion that it understands that which is beyond human comprehension.

(b) Analogies are never 'true'. They all fail ultimately as they are put to the test in new situations, but then, surely, so do all the so-called laws of physics.

(c) An analogy which is useful for one person may detract from the performance of another. 'Teaching' (or what is more dogmatic, 'preaching') of analogues should therefore always be done in the context of 'take it or leave it'.

(d) Analogy should be, despite all the foregoing pitfalls, the daily staple diet of the would-be inventor.

The process of analogy is most easily explainable in terms of examples. We can begin with a relatively simple one. A man whose London office window looked out over Parliament Square and the great clock tower of Big Ben, but who rang the telephone exchange daily to find out what time it was, could be accused of 'Having a dog and barking yourself'. Analogue is used to emphasise a point, in this case to ridicule. In defence of a policy which involves some unpleasant acts, a phrase often attributed to Mr Kruschev is that 'you don't make an omelette without cracking eggs'. In this example the analogue is used to clarify something which might otherwise be obscure. Shakespeare made frequent use of similar analogues: 'Gratiano speaks an infinite deal of nothing, more than any man in all Venice. His reasons are as two grains of wheat hid in two bushels of chaff: you shall seek all day ere you find them; and, when you have them, they are not worth the search.'

Pursuing the idea to a more advanced form and nearer to what the scientist needs, we can find excellent examples in the New Testament of the Bible, when Jesus's many parables were intended to describe that which was beyond the human mind, in terms which would enable his audience to get an appreciation of at least one aspect of the subject: 'The Kingdom of Heaven is like unto a grain of mustard seed . . .' And of course, he needed more than *one* of these parables for a subject as demanding as comprehension of 'the Kingdom of Heaven'. These are analogues of use rather than of wit, as were the simple ones I have just given. But of course the use of analogue is a good deal older than 2000 years. It is a part of Man's make-up and therefore as old as human intelligence.

One of the most useful ways of conceiving analogues is by recognition of an *order* or *pattern* which arises in one field of interest but which is familiar to us in an entirely *different* type of situation. Often the order or pattern is seen most readily in a mathematical equation, for mathematics is to the engineer little more than shorthand is to the stenographer. It is a concise way of writing in a single line that which would occupy a page if expressed in words. Of course, such shorthand lends itself much more readily to 'organisation' than its lengthy word equivalent.

There are two major aspects of engineering, namely, *hardware* and *concepts*. One may invent either or both. An engineer seeks always to pursue those things which are *more profitable*, either in terms of hardware (£s) or concepts (Fellowships of learned societies), the rewards for each being shown in brackets!

A vital ingredient which applies not only to invention but to problem solving in general is the ability to strip a problem or situation of all its trivia and relatively unimportant detail and concentrate only on the very bedrock of the system, an ingredient generally born of experience alone. To teach problem solving is more difficult than teaching certain aspects of invention. It has been for many years a source of frequent frustration for me that whilst I can look at a school 'A' level paper in Applied Mathematics and within a few seconds 'see through' any given question (by 'see through' I mean know exactly how to solve the problem in the quickest way, see all the steps involved and find no more difficulty in starting than I have in starting to eat a mixed grill!), yet I cannot teach an eighteen-year-old pupil this ability. I can only hope that the technique emerges for them by constant practice in problem solving. So also do I see the same situation in one aspect of invention. Where the invention *begins* with a problem this is especially relevant. Where invention begins with *observation* (as have 90 per cent of my own inventions) experience is helpful in rapid exploitation of the phenomenon, but never vital. Where curiosity is the spur, the end is seldom obscure. Where help in a problem situation is demanded, the would-be inventor is much more like a man handcuffed to a wall, trying to write a letter using his feet.

SYMMETRY

There is within us a built-in desire for symmetry. In symmetry we see a beauty which cannot be set down in equations – it lasts perhaps no longer than an instant, but it has influenced our engineering, our buildings and our cities, as well as our machines and our gadgets. It is not an easy thing to define in a few words. Possibly it is best to illustrate it by means of a few examples. A spade is symmetrical about a plane at right-angles to the plane of figure 5.1(a) that cuts it in half from top to bottom, but a similarly placed plane to cut figure 5.1(b) will not divide it symmetrically. Perhaps then we

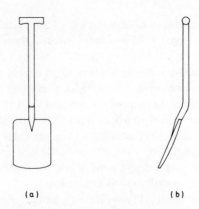

Figure 5.1   Symmetry; a spade is symmetrical about one plane (a), *but not about a plane that lies at right-angles* (b)

should start simply with two-dimensional pictures. The pear shape shown in figure 5.2(a) is symmetrical about the line AB, because for *every* point such as X on the outline left of AB there exists a corresponding point Y which is exactly as far from the line AB on the right as X was on the left. The line AB is called the 'axis of symmetry'. The shape shown in figure 5.2(b) has no axis of symmetry and so is an asymmetric shape. Real pears are always like (b), never like (a), but we nevertheless often think of a 'perfect' shape such as (a) as being 'like a pear'.

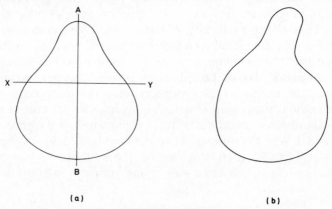

(a)   (b)

Figure 5.2   Symmetry and asymmetry. (a) *The pear shape is symmetrical about axis AB;* (b) *a real pear is never symmetrical*

Even two-dimensional objects can have more than one axis of symmetry. The rectangle shown in figure 5.3 is symmetrical about both AB and CD. It is *not* symmetrical about a diagonal, but a square *is* (see figure 5.4(a)). A circle has an infinite number of axes of symmetry, whereas an ellipse has only two. Thinking out why this is and where any one axis is, can be, if you like, the *beginning* of physical thinking.

72

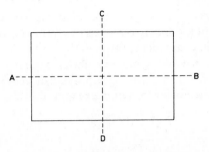

Figure 5.3   Object with two axes of symmetry

A good test for an axis of symmetry in a flat picture is to place a mirror with its plane perpendicular to the picture, along the line to be tested (AA in figure 5.4(b)). Looking into the mirror should give the impression of the complete picture as if no mirror were present. Any plane figure with at least one axis of symmetry can be so placed and is said to be 'symmetric', although there are other, more abstract definitions of symmetry, such as those used in higher mathematics.

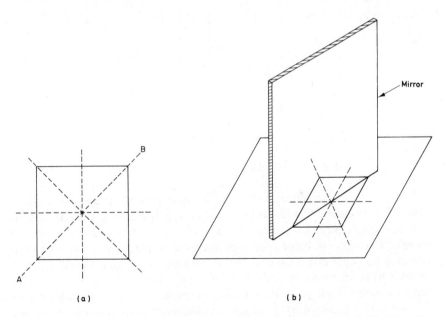

(a)   (b)

Figure 5.4   (a) A square is also symmetrical about a diagonal. (b) Use of a mirror to test for axes of symmetry

In three dimensions, there can be *planes* of symmetry or axes of symmetry, or both. The cylinder shown in figure 5.5 is symmetrical about the axis PQ and also about any *plane* that contains PQ – one axis, an infinity of planes. It has an additional plane of symmetry about its mid section, as shown

73

dotted. A sphere has an infinity of *axes* of symmetry, i.e. *any* straight line through the centre of the sphere is an axis of symmetry. It also has an infinity of *planes* of symmetry (any plane through the centre of the sphere). It therefore has three-dimensional symmetry about a *point*, i.e. its centre. 3-D symmetry is rare, even in engineering. If we return to our spade in figure 5.1, it can be seen to have no axis of symmetry and only *one* plane of symmetry.

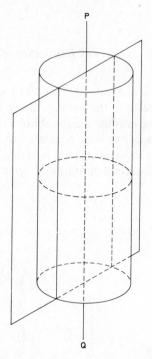

Figure 5.5   Axes of symmetry in a cylinder

Test yourselves in physical thinking by seeing whether you can confirm mentally that a cube has 13 axes of symmetry, 9 planes of symmetry and no 3-dimensional (point) symmetry. When you have mastered that exercise imagine that you see before you a cube that is so placed that, with one eye closed, your other eye is looking directly along a major diagonal of the cube (were it to be transparent, the corner that points at you would obscure the furthest corner from you). If however, the cube is solid, what shape would you see (still with one eye) in outline? In other words, what shape of shadow would it cast if illuminated by a point source of light at the position of your eye?

The answer is shown in figure 5.6, deliberately placed on the adjacent page, in case you want to wrestle with the problem without cheating. Most people, as soon as the word 'cube' is mentioned, think of 'square' and the idea of 'square' implants the number 4 in their heads, so that most answers are 'a diamond-shape', or 'a square'. The number that should dominate for a good physical thinker is 3, not 4, and the outline is a hexagon. What is more it is regular. Even when a regular hexagon is drawn as in figure 5.6(a), some people still cannot see it as a cube in outline, until the extra three lines are put in, as in figure 5.6(b).

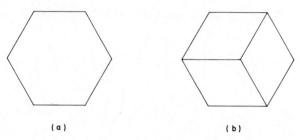

Figure 5.6   Regular hexagon (a) is the outline of a 'point' view (b) of a cube

Perhaps now it is more obvious why symmetry is a difficult thing to define. There are different kinds of it, some relating to flat things, others to solid things. In the case of solids there are several kinds of axes of symmetry. A cube can be rotated about the line MN in figure 5.7(a), can be set in any one of 4 positions and all are indistinguishable from each other by an outside observer. MN is called a 'fourfold' axis of symmetry, and a cube has three of them. But the cube in figure 5.7(b) can only be rotated to one of two identical positions about OP, so this is a 'twofold' axis of symmetry and a cube has six

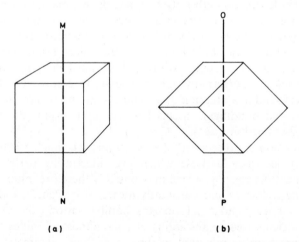

Figure 5.7   Axes of symmetry in solids. In (a) MN is a 'fourfold' axis of symmetry; in (b) OP is a 'twofold' axis of symmetry

of them. Crystals can have axes of twofold, threefold, fourfold and sixfold symmetry, but never fivefold. The cube has four axes of threefold symmetry, one of which is shown in figure 5.8.

75

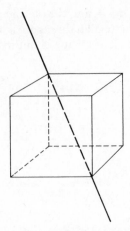

Figure 5.8   An axis of symmetry in a cube; a cube has four axes of threefold symmetry

A teacup with a handle has only one plane of symmetry – hence the joke about the 'left-handed teacups'. But it is well to remember that there are three dimensions known to us and a fourth that hovers always out of our grasp. There are examples of each to be found both in nature and in man-made objects. Generally those with only one axis of symmetry are more common than those with two, those with two more common than those with three – and what of those with symmetry in the fourth?*

Of man-made objects with only one axis or one plane of symmetry like our spade, we can think of knives and forks, pen nibs, zip fasteners, cars, ships, aeroplanes (the last three in superficial outward appearance only), bicycles, an assortment of clothing (apart from decoration and the fastenings), the majority of weapons (swords, clubs, guns, etc.) and some houses and other buildings, externally only of course. In nature, one-dimensional symmetry is found in all mammals and birds (externally only of course, like the vehicles and buildings), insects likewise, and certain stages in plants (e.g. at the 2-cotyledon stage).

The proportion of objects with 2-axis symmetry is much greater in manufactured articles than in those which live – illustrating perhaps our preoccupation with symmetry, which may possibly therefore inhibit our inventive thinking unless we are constantly aware of it. Pencils and ball-point pens, wheels, bottles and jars, food cans, bombs, buttons, the essential parts of most electric motors,† nails and rivets, and almost all things made out of tubes. Nature is more reluctant in the use of 2-axis symmetry, but many flowers and stems have at least a close approximation to circular form (and

*It will form an interesting test for the reader at this point to attempt a definition of symmetry in a fourth dimension. Some of our ideas will be disclosed later in this chapter, but not here so as not to influence your imagination, for we would guess that the idea of symmetry in a fourth dimension is new to most of you and therefore a good test of your imaginative powers.
†But see what more can be done when 2-dimensional symmetry is abandoned in electric motors: Laithwaite, E. R., *A history of linear electric motors*, Peter Peregrinus (1976).

therefore to 2-axis symmetry). Botanical life in fact uses it much more than does zoological.

Three-dimensional symmetry does not imply spheres alone, but anything based upon spherical symmetry, so man decorates his buildings, his clothing and draws his pictures on the inspiration of the sphere; he of course worships the sphere on a Saturday when the position of inflated leather spheres in various fields throughout the country spells glory or ignominy for the few who chase it, happiness or gloom for those who watch them do it, and riches beyond belief for those who predict where eight particular spheres will find themselves on each successive Saturday.*

Nature usually contents herself with the eggs and seeds – the beginnings of life – for spherically symmetrical objects, and there may be clues for us here as inventors if only our biological knowledge extended far enough, for it is in nature that we find a far greater proliferation of completely asymmetrical objects. Nature, after all, is not ruled by the accountant and gentlemen of that profession should be restrained from seeing the Creator as 'the greatest of all accountants', even if we agree that all things in nature are profitable in the long term, for what accountant would allow 5000 men to begin carving a statue to a famous man on the grounds that 4999 of them would shatter their stone in the process and thus only one perfect object emerge? (c.f. the eggs of insects, spores of fungi, animal sperms.)

An interesting variant on the two-dimensionally symmetrical objects is the object which is tubular or conical or pyramid-shaped, with an infrastructure built into it which gives it a left- and right-handedness – to our way of thinking. Simple examples are seashells (natural) and screws (manufactured). The very concept of left and right has its origin in symmetry. When we look in a mirror our left hand becomes our right, so why do not our head and feet also change places? Why does a clockwise rotating wheel appear to rotate anti-clockwise in a mirror? The answer to these far from simple questions lies in the inversion *front-to-back* which a mirror performs.[1] Had we, the observers, had symmetry in two dimensions (we would have had perhaps an eye on each side of our heads, a mouth in the centre of the top of our head, a nose on diametrically opposite sides midway between the eyes, etc.), we would not have recognised a left and right in ourselves, yet we would have been able to distinguish a wheel in rotation as 'different' from its image in a mirror. It follows that an object with 2-dimensional symmetry loses a dimension if it elects to spin about the axis which is common to those two dimensions.

Mathematicians have developed their talents until they can work entirely in the abstract and therein they find an elegance, a beauty (a part of which is symmetry) which rivals that of the artistic eye for a Gainsborough or a Van Gogh. Sir James Jeans once declared 'God is a mathematician.' This is

---

*Rugby footballs, of course, have 3-axis symmetry which is not spherical. In *Brave New World* Aldous Huxley has his humans of the future play tennis on Cauchy-Riemann surfaces rather than horizontal planes. It seems highly doubtful that this will ever happen in reality.

laudable although never to be translated by lesser men into 'I, a mathematician, am God-like'!

THE 4–D MIND

What then of symmetry in the fourth dimension? One simple concept is that a symmetrical occurrence in time is one which, if recorded on ciné film, appears identical, whether the film be run forwards or backwards. How many events can you name which fit this definition (for this is all that it is) of time symmetry? One or two examples only, just to start your thinking – anything with repetitive oscillatory motion of constant amplitude. This example needs to be taken with caution, for most *natural* oscillations are inherently damped (e.g. a simple pendulum) and your eye will detect the decay or build-up, depending on the film direction. A child on a swing can maintain a constant amplitude (or even a variable one), but it *must* do so by an essentially asymmetric motion which our eye could learn to detect. Symmetry to the rescue again! But the valve rockers of a car engine would be symmetrical. Similarly a person scratching their face is undetectable, provided that the start and finish of the operation is in a random position. The view of a person's face whilst that person is speaking is, to the untrained eye, symmetrical. But a good lip reader can easily distinguish forwards from backwards.

If purely oscillatory motion (and for simple equations of motion only) is time symmetrical, so also must be rotation, up to a point. If the rotation be slow enough for the observer to identify the parts revolving, and the parts themselves be space asymmetric (for example, the blades of a fan intended to drive air in a known direction), then the motion is not time symmetric.

The late Sir Lawrence Bragg put together some magnificent words on the subject of past and future:

'When we postulate an experimental set-up, and wish to prophesy the result, we must treat both light and matter as waves. The nature of physical reality is such that we can only calculate the relative probability of an effect in various places. On the other hand, when we write the history of what did actually happen in an experiment, it is a history of particles whether of matter or light. A wave-like uncertain future, only expressible in probabilities, forever streaming through the moment "Now", is transformed into a definite past of particles. Determinism takes on a new meaning.'

This implies that time is basically asymmetric in the human mind. If so shall we ever be good enough to escape from this mental prison? (Of which there are many!) One could even express human ignorance as a series of walled prisons, one inside the other. From time to time throughout the 'Ascent of Man', to use Dr Bronowski's words, we have broken free from one prison, only to find another wall outside, but nevertheless the area over which we have free range gets bigger with each successive break-out.

78

Time symmetry is a great brain-stretching exercise for the inventors of tomorrow.

REFERENCE

1. Laithwaite, E. R., *Engineering Through the Looking Glass*, B.B.C. Publications (1976)

6 ENGINEERING AND NATURE STUDY

E. R. Laithwaite

It has often been said that there is 'nothing new under the sun', and what is usually meant is that the *principle* is not new, even if the embodiment (to use a patent agent's phrase) is novel. A simple example is a ball and socket joint which I am sure was invented without reference to the bone structures of the limbs of animals. Gilbert Walton writes[1] that such well-used phenomena as those of steam engines and electric currents were abundant in nature long before man was aware of them, but that in the one case they were too small and in the other, too large, for us to notice them. Too small, we can understand. The currents to which he referred were those associated with nerve impulses and our science had to attain a high degree of sophistication before we could possibly become aware of them. But too big! How could we fail to observe something because it was too big? He referred to the vast amounts of water evaporated daily from the sea and deposited high up on the land where its potential energy is available to drive waterwheels and turbines. It is a world-sized steam engine complete with boiler and although the water does not boil, the principle is the same.

There is a long list of inventions that man claims to have made, each of which was subsequently shown to have existed for millions of years earlier in living creatures. Somehow we have to discover these things for ourselves before we notice them in our nature study – all of which shows how very bad we are at nature study. It has been said unkindly of biologists that their motto is 'observe all things, record all things, classify all things, but may God preserve us from ever making a deduction'. Biologists however seem to be better equipped for making original observations than the rest of us.

'IT WAS THERE ALL THE TIME'

Let us examine some of nature's engineering feats in the hope that 'the penny will drop' and that we might begin to make better use of the common things we see around us. We may begin with what appears to be so closely linked with the wheel that natural things might be assumed to have no need of it. I refer to the principle of the underslung chassis, whereby the designer of carriages ensures that the centre of gravity of the vehicle is lower than the points from which it is suspended. But look at figure 6.1 for an elaborate set of legs with knee-joints well above the body mass. Spiders, beetles, flies – a great variety of creatures use this stabilising system to good effect. For some species it is as vital to life as are the lungs of animals, for if the insect is ever to fall on his back, it cannot right itself and dies there, slowly.

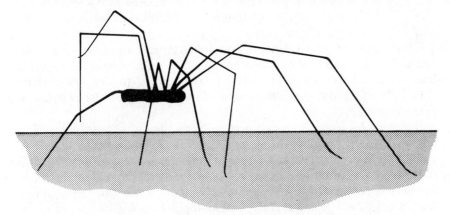

Figure 6.1   An insect with an underslung chassis

Anyone who has been stung by a bee or wasp needs no reminder of nature's hypodermic needle, used by plants and insects alike.

The use of tools is often quoted as the demarcation line between 'intelligent' man and 'instinctive' animal. In particular our ingenuity in catching fish by lures and baits stretches back propably into prehistory. But it is so obvious. Most other creatures have their hunting equipment built into the system, so that it is self-healing and in many cases can be regrown if damaged. It requires no maintenance, as does our fishing tackle, and if we could obtain the average sea creature's views on the angler fish that dangles out a piece of its own body to act as bait and is to be seen as the true 'fisherman', we should probably be told that this creature was regarded as 'primitive' rather than sophisticated in its methods.

HIGH TECHNOLOGY

The further we advance in technology, the more of it we seem to be able to identify in living creatures. Surely we must reach a stage beyond which we surpass all of the 'lower animals' in technique? Apparently not, at least not up until now.

We are proud of our high-tensile steel, but it is inferior in strength/weight ratio to the silk of the spider's web. We reckon ourselves good at accurate measurement of physical quantities, yet when we make a dark room and need proof of its 'total darkness' we resort to the use of a 'vegetable' to detect the odd photon that gets through. Two American authors, writing in the American Institute of Electrical and Electronics Engineers journal *Spectrum*,[2] describe this sporangiophore which turns its 'flower' towards the light 'leak' in a matter of hours. If we could reproduce the detector, doubtless we would do so. The same authors have this to say about our adhesives: 'If one could discover the nature of the glue with which barnacles bind themselves so avidly to surfaces...and bottle this substance, one might reap a fortune.' The glow-worm's light is 'cold light', i.e. not associated with a temperature rise. That it is some kind of 'oxidation process' is about as far as our knowledge takes us in this phenomenon.

The common bedbug detects his 'host' by a temperature difference between one end of his proboscis and the other at a distance of several feet from the human body. This involves measuring temperature to an accuracy of better than $1/1000°$C. Snakes are also capable of measuring this order of temperature difference.

Biologists have sometimes been as guilty as the rest of us in refusing to admit that there are phenomena we do not understand. If for example an animal succeeds in detecting an object or another living creature when it is obvious that it cannot hear or see it, the usual refuge is to declare the mechanism to be 'of an olfactory nature' which is a good scientific way of saying that 'it smells it'. As an amateur entomologist I read with interest in the 1940s that the males of a certain species of moth could locate a female 'at a distance of a mile'. Whilst test-flying automatic pilots, I had time to spare and noticed how smoke from factory chimneys and from railway engines was soon broken up into eddies and whirlpools so that no ordinary sense of smell could possibly indicate the source, for to obtain a sense of direction from a particulate smoke requires the following two features to coexist.

(i) A relative movement between source and observer. (Watch a dog moving continuously in a left-right direction when tracking an animal.)

(ii) The ability to 'remember' earlier strengths of smell. This could be interpreted mathematically as the ability to take a time derivative of intensity.

The story of my search for the true long-range detector of the moth is a long one, and is recorded elsewhere,[3] but it took me into the subjects of black-body radiation, diffraction gratings, microwaves, aerials and resonant cavities as found in magnetrons – but *nothing new to physics*. A man more skilled than I in biological techniques finally located the radiation receptors in the antennae of the moth[4] (why else were the receptors of radar called 'antennae'?) but I doubt if his theories would have been as readily acceptable had we not already developed our own radar techniques.

Let us for a moment go back in time to the 1790s, when a biologist named Spallanzani had done experiments on bats to investigate their ability to fly in total darkness. He had blocked their throats and showed that when unable to cry out, they could apparently no longer 'see'. But he was discredited. Experiments involving any form of mutilation were taboo. Perhaps the worst that happened to poor Spallanzani was that one of the accepted authorities of that time, in debate before a learned society, asked of him, 'If they see with their ears, what do they hear with their eyes?' and the hall rocked with derisory laughter. But in this enlightened century, we might well comment in all seriousness: 'What a *good* question!' This is precisely the sort of question that leads to good invention. In respect of the bat, of course, we had to wait over a century before we really *knew*, and we only knew after we had invented the system for ourselves. If nature had developed the internal combustion engine 100000 years ago, we should not have been able to copy it until modern times for we would not have known what that 'sack of juice' was for, since we would not have been able to refine crude oil.

Some thirty years after the bat's 'visual' mechanism was accepted as acoustic, Dennis Gabor invented holography and a further decade was to elapse before the laser hit the headlines and put holography on a commercial footing. Within a year a zoologist proclaimed that the acoustic radar was holographic. 'Of course,' we all said, 'the bat wishes to see in three dimensions as do we ourselves.'

There is another aspect of human invention that has a certain similarity to the 'of course, there it was all the time' phenomenon of observing in nature what we have just invented, and this can probably be described best by means of an example. If alternating current technology had persisted after Pixii's alternator of 1831 and he had listened more to Faraday than to Ampère, the induction motor might have come first. If then, in 1974, a man had come up with the idea of the commutator and brushes as a means of

(a) driving motors from batteries,
(b) adjusting the speed of a.c. motors without incurring extra loss,
(c) improving the power factor of induction motors,

I can tell you what would most likely have happened to that perfectly good invention. He would have been told that:

(1) Industry did not want it.
(2) N.R.D.C. could not afford to develop it.
(3) It would only be useful possibly for toys, since the commutator bars could not be embedded sufficiently firmly to resist centrifugal force.
(4) It would be too expensive, even for toys.

Wasn't it lucky the commutator got in first, for we still find it a most useful device!

The processes of evolution that have developed the life forms we know appear almost to have hidden their secrets deliberately to force *homo sapiens* into a gigantic hide-and-seek game, the whodunnit we call 'science'. One of the favourite concealments that appears to be aimed at our confusion is to use the same external organs for more than one purpose. The knee-joints of some insects carry 'ears'. In the grasshopper family the legs are used as 'voices'.

The value of nature study to the inventor was evident to Leonardo da

Vinci, and if it was good enough for Leonardo it will surely help us also. We need to look at the shapes in nature and ask ourselves their purpose. Even when we *know* the purpose of a particular piece of living tissue, we need to look again and try to find a second use, almost with the same spirit with which an effective shop assistant tries always for the second sale, to a customer who apparently had entered the shop with but a single purpose.

Figure 6.2 is a picture of something most of us would admit had a certain shapeliness that might be admired as decorative. But figure 6.2 is a picture of dung – the dung of the Viennese Emperor moth. Only when immersed in water does the dung separate into these laminae. On the basis that many things that look attractive are also highly functional, what is the purpose of this most elaborate form of waste matter? Surely a dual-purpose instrument created this, like sugar icing flowers emerge from the nozzle of a pastrycook's forcing bag, but why that particular shape of nozzle?

Figure 6.2   Laminar dung of the Viennese Emperor moth

With most birds' eggs, surely the answer to the question of their shape must be to ease open the egg vent gradually with the sharp end first (see figure 6.3)? Not a bit of it, the egg comes out blunt end first. I am indebted to my fellow engineer and friend, Jack Ferguson, for writing to me on the subject of the eggs of the guillemot after hearing my comments on the shape of eggs. He writes: 'Guillemots lay their eggs on ledges of cliffs, generally in fairly inaccessible places. high up and subject to strong winds. The nests are non-existent or very rudimentary. If the egg rolls at all, it rolls around in tight circles and does not run off the ledge.

'In the old days when there used to be fishermen sailing out of Flamborough (Flamborough Head used to be a great nesting area for guillemots) I have heard them swear that they have seen guillemot's eggs spinning round like

Figure 6.3   A guillemot's egg

tops in the high winds, but staying on the narrow ledges where they had been laid.'

Of course the smooth surface and general shape of the egg are also helpful to the laying bird. It is a dual-purpose shape and, of course, it is visually attractive.

'BETTER THAN WE ARE'

An inventor must have versatility in all its facets. He must sometimes be stubborn, at others most flexible. He must be capable of believing in himself and his ideas almost to the state of arrogance. But he must also know a humility of great depth, especially when he looks to natural things for inspiration. For there is an abundance of activities practised by the lower animals, many of which leave our own technology almost at the starting gate. The flight of birds and insects is an obvious example, but of course we start with a crippling handicap of being already bigger than the natural flight size governed by the rules of planet Earth and we could argue that we fare much better in water than do butterflies or earwigs.

But in a subject common to most living things, the best example being communication, one with another of the same species, can we be *sure* that we are winning? For most of the time we do it in serial time. In writing, sentence must follow sentence, in a logical sequence. Within a sentence, word must follow word, and comply with the most complex of rules – grammar and idiom we call them, and the rules are very different in different languages. Even in a word, the letters must be in the right order, and not even letters are the same in the written forms of communication of the geographical races of man. The effect of the Tower of Babel has persisted thus far.

Telepathic communication is on the brink of becoming scientifically respectable, but is this a dying sense or a new one, and is it practised by any other creature? I have seen a flock of starlings, perhaps several thousand in number, make a sharp-angled turn in flight. They turned like a regiment of well-drilled soldiers, simultaneously as if on a word of command, but so far as we know, there *was* no word of command – unless it was silent!

Television is a wonderful creation of civilised man, but just how good is it as a means of communication? In order to make it work we divide the picture in the vertical into hundreds of individual lines and proceed to make a spot of light appear to travel along each line in turn, following very closely the kind of path on the screen that my pen follows on this page that I now write, except that the *signal* is a modulation of light intensity rather than of small modulations of vertical and horizontal movement. So we must repeat the picture sufficiently frequently to convince the human eye that there is no moving spot, no lines, no flicker and this fixes the minimum possible line-scan frequency; the latter in turn fixes the range of signal frequencies, and the transmission 'carrier' frequency must be many times the highest signal frequency. When all this has been worked out it emerges that the transmission frequency is too high to be bounced off the upper atmosphere as in radio and that unless each receiver is virtually within visual range of the transmitter, good reception is not possible.

Now we often think of nature as being extremely wasteful in her methods, especially when it comes to spores, sperms and seeds, for to preserve the balance of nature, only one adult female must result from hundreds, thousands, sometimes millions of these 'life transmitters'. But let us turn the same critical spotlight on to our revered television system and see just what we have achieved. We can, if we wish, consider each line of a TV picture to be capable of subdivision into perhaps 1000 'bits', so that the whole picture is effectively made up of 625000 bits of information. The system is capable of changing each one of these 25 times per second, but we *never* use this facility. Our maximum demand on the system occurs when the production director switches from one camera to the next to give a complete change of scene. But I am sure we would settle for a gradual fade and reappearance which could occupy perhaps $\frac{1}{4}$ of a second. If the whole scene were to change 25 times a second we should receive far less intelligent information than we would about a hen's egg from a superficial visual examination of scrambled egg!

In most plays, documentaries, and the like, the vast majority (well over 99 per cent) of the 625000 bits remain unchanged for long periods. Only the cartoonist knows how to capitalise on this fact. If we could exploit it electronically we could reduce the bandwidth and make the transmission of TV signals as easy as radio. We could scan the world without artificial satellites and reduce the costs of both transmission and receiving sets by a huge proportion. Let us not be complacent about our high technology. There is room for vast improvements in human communication of all kinds. See how many pages Sir Walter Scott needed to describe the dining hall of Cedric the Saxon in *Ivanhoe* and how long it takes us to read it. The same amount of information could be obtained by a well-trained eye in perhaps no more than a second for the eye does not rely entirely upon scan, and makes much use of familiar objects which require no detailed examination as is necessary in the written or spoken method of description.

A 'broader education' such as many strive to achieve today could have been the end of fertility for the brain of a modern Newton, Faraday or Maxwell at the tender age of about thirteen. We have no real formula for the betterment of education of today's children, I suggest, unless a particular class of children has the benefit of a dedicated teacher who has enthusiasm of such calibre that he induces in each one the *will to learn*. After *that* seed has been planted, it will grow and flourish in almost any soil. Such a teacher can at one and the same time excite a pupil yet give him a sense of true humility, and biology is surely one of the recipes. We know all about hexagonal structures from an engineering viewpoint. Scientifically we know about valves and siphons. Industrially we are proud of our zip fasteners and our Velchros that hold pieces of material together at a touch. So who told the honeybee about the first of these, the inhabitants of seashells about the second and third, and certain species of moth about Velchro?[5]

In the year of the inaugural conference of the world's first full-scale computer (1951), the acknowledged mathematical genius, the late Alan M. Turing, whom I am privileged to have known, made an estimation of the size of machine that would be necessary to ensure that it succeeded in its only purpose – to create other machines just like itself. It must obtain its own ingredients, use only solar energy, and be restricted by 1951 technology. He reached the conclusion that the minimum volume of machine necessary would be approximately one-half the volume of the earth. On the next occasion when you are lying in the sunshine and are aware of a tiny black speck (that is probably an insect) crawling over the back of your hand, reflect that this tiny creature can perform the task that Alan Turing set his machine quite easily. We have at least *that* far to go!

I contemplated an example of the same kind of exercise in a lower key by supposing that a government hand-picked a half dozen top scientists and isolated them and their families from communication with the outside world, but gave them all the comforts and scientific equipment they required (except radio receivers and transmitters, or the means to build them!) on the understanding that in two years' time, for the benefit of mankind in these days of energy crisis they should emerge with a device that would pump water from the ground continuously, from an area of 50 square metres and a depth of 2 metres (100 cubic metres of earth weighing over 100 tonnes), and should use no power supply other than that obtainable directly from the sun, and should go on working unattended for over 100 years. I doubt very much whether, in the technologically enlightened year of 1976, they would emerge with anything that looked, even remotely, like a tree.

Think about it!

REFERENCES

1. Walton, G., 'Facts and Artefacts', *The Modern Churchman*, **3**, 233–238, July (1964)
2. Gamow, R. I. and Harris, J. F., 'What Engineers can Learn from Nature', *Spectrum*, **9** No. 8, 36–42, Aug. (1972)
3. Laithwaite, E. R., 'The Magnetic Butterfly', *Proc. Roy. Instn.*, **46**, 1–17 (1973)
4. Callahan, P. S., 'Insect Behaviour', Four Winds Press, New York (1970)
5. Laithwaite, E. R., Watson, A. and Whalley, P. E. S., *The Dictionary of Butterflies and Moths in Colour*, Michael Joseph (1975)

7 THINKING WITH THE HANDS

M. W. Thring

SOME WELL-KNOWN EXAMPLES

In chapter 1 the theory has been developed that invention, like any other creative art, requires the working together of a man's three brains. The need for its use of the physical brain or 'thinking with the hands' is not nearly so widely appreciated as is the use of the intellectual brain. Even Archimedes, who seems to have suffered from the Greek contempt for the mechanical arts, used his own experimental observation that when he got into a full bath the water overflowed, together with his reasoning that the amount displaced had a volume equal to that of the displacing body, to invent a solution to the problem of determining the density of the metal of a complex shaped crown. His emotional brain was clearly involved which is why he rushed home out of the bath shouting '*Eureka*'.

In this chapter I shall first try to demonstrate, by examples, how various inventors made use of their ability to think with the hands and then suggest some ways of improving this ability in the course of a person's education. The aim is clearly that this ability shall be as fully developed by appropriate exercises as the intellectual ability is meant to be developed by passing examinations.

Thinking with the hands can only be produced by a deliberate combination of constructing one's own experimental apparatus or working models and then observing carefully what happens. Jack Chesters described in his Presidential Address to the Iron and Steel Institute how he familiarised himself with the behaviour of eddies in turbulent flow in irregular shapes by putting sawdust on water in the kitchen sink and watching water flow past bridges. C. V. Boys spent many years inventing and developing very accurate calorimeters to give the calorific value of towns' gas and other combustible gases, with automatic compensation for conditions of varying temperature, pressure and humidity. He constructed all his own models because he found that this was essential to bring his ideas into shape as the constructional work proceeded; the work helped the thinking process and he would not get on quicker by having it done for him. Buckminster Fuller who has produced so many new inventions in the field of static structures was brought up on a small island where sailing ships and their highly developed structures of ropes and spars were part of his daily physical education. There is no doubt that his exceptional ability to think in three dimensions and visualise the forces leading to static stability gave him the necessary freedom to imagine entirely new ways of supporting objects in space and to design on the surface of a sphere.

A much earlier example of thinking with the hands is provided by the skilled fletcher who knew when an arrow had the right properties by flexing it. He did not know the theory that it had to have such a natural frequency of flexural vibration that the tail would have carried out exactly 180° of oscillation by the time it passed the bow, so that from being bent towards the bow by the string, it was now fully clear of it. Any skilled craftsman who shapes the handle of a tool until it 'feels right' is using a similar ability, although by now most of the best shapes for hand-tool handles have been fully worked out by such empirical means. Anyone who has found how easily an axe handle splits if the back of the axe is used as a sledge-hammer will appreciate the beauty of the design of the handle for its proper purpose. The shape of an adze so that a groove can be cut out of the wood is only appreciated by using one. The wood carver always has a favourite tool which he has reground to suit his ideas.

WHY THE ABILITY TO 'THINK WITH THE HANDS' IS ESSENTIAL TO THE INVENTOR

In the case of a static structure like Fuller's geodesic dome or of the design of a simple mechanism like a gripping apple-picker, it is obvious how the inventive act takes place, primarily through one's appreciation of reality with the hands. It is rather more difficult to see how Edison's 'intuition' enabled him to visualise that high-resistance lamps in parallel would enable electric lighting to be operated remote from the dynamo, without using excessively heavy copper lead wires. In fact it is certain that his hunch came from the knowledge of electric currents that he had obtained by experimenting with the telegraph, with batteries and with arc lights, and it is this kind of practical knowledge that we mean by 'thinking with the hands'.

This ability is essential to all the four stages of any invention that involves working hardware or physical reality; the only kind of invention that can be done by a purely intellectual process is one in the realms of pure mathematics. The first stage is the analysis of the problem in the simplest possible and least restrictive terms and the picking out of the centre of gravity or key difficulty. This process might at first sight be thought to be purely intellectual like solving a mathematical puzzle, but in fact it requires the emotional brain to make the judgment of what it is that one really wants to achieve as well as the physical brain. It requires the physical brain or knowledge with the hands because all intellectual knowledge is essentially abstracted from the reality of objects in space–time.

The intellectual brain can only deal with concepts, words, ideas and theories, whereas it is through the senses and the hands that one has experience of the behaviour of solids, fluids, flames, explosions, mechanisms, electric currents, magnetism, forces and heat. These I shall refer to as 'hardware' to distinguish them from the abstractions used by the intellectual brain. From this experience one can construct concepts and theories, but any teacher of physics knows that a student who has had no contact with experimental observation is totally limited.

The role of the physical brain in the understanding of the problem of what one is trying to invent is therefore to make sure that one is grasping a real space–time problem and not a theoretical problem existing in logic alone. As far as possible therefore one prepares oneself for the inventive act by playing with models, sketching diagrams and looking at the working of existing solutions to the problem. One gets as much physical knowledge as one can of the problem and of the magnitude of the forces, velocities, temperatures and pressures involved. Crude models roughly put together in one's own work-shop can play a very useful role in giving one's hands the ability to think about the problem.

The physical brain enters into the actual inventive act as an active partner with the intellectual and emotional brains. Its role is to connect the vague concept produced by the intellectual brain with reality, to clothe it in hardware (taken in the broadest possible sense as defined above, to include all measurable and sensory perceptive objects) so that it can have potential existence in real space–time. Sometimes it produces the new shape or arrangement before the intellectual brain. When one has produced this potential invention, one comes to the third stage which is to bring it to the point where a working model or first prototype can be constructed for experimental testing. Here the physical brain plays the most important role. Rough sketches are the first technique. The back of an envelope is the most expressive description of the medium used, but a pocket notebook is the inventor's usual companion, preferably of unlined blank paper, and the sketches are made very roughly so that one has no compunction in throwing dozens away while the idea is being tested and taking shape. The ability to draw effectively freehand is just as essential a part of the inventor's tool-kit as it is for the lecturer and the artist. Drawing with a soft pencil on unlined paper enables one to use a rubber frequently, which is a valuable aid to changing one's ideas. One frequently needs to rub out lines that are hidden behind other parts.

Sketches are very limited by the fact that one cannot readily visualise three-dimensional situations – for example one cannot draw Fuller's geo-desics on anything but a sphere to see how he achieves approximations to geodesics by a combination of six-sided and five-sided polygons. Similarly one cannot readily visualise the movement of double cranks so that they do not foul one another – a problem I had considerable difficulty with in developing the Centipede (see chapter 10). Mechanisms involving move-ments of complex shapes, for example the rotary engine, can well justify making models with cardboard cut-outs and using drawing pins for rotation axes.

The intellectual brain is involved here mainly in solving problems thrown up by the attempt to work out the idea physically.

In the fourth or development stage, which is dealt with fully in chapter 9, the ability to think with the hands is vital, both to ensure that design calculations produce sensible results, and that all the numerous parts that cannot be calculated 'feel right' or 'make sense'. Every good designer has acquired the ability to look at a drawing and spot dangers which will arise because of errors or lack of understanding of the problem. None of the phrases 'feeling', 'making sense', 'understanding' is quite right for this activity of the physical brain which we have called 'thinking with the hands'; the words *manualising* and *dexterising* have however been suggested.

TRAINING THE ABILITY TO VISUALISE IN SPACE–TIME

The inventor must acquire the ability to visualise complex static systems in three-dimensional space and to relate two-dimensional sketches to real systems. The classical lesson of constructing Moebius bands (figure 7.1(a)) out of paper strips with the ends gummed together after 180° and 360° twists, and then cutting them longitudinally, and the more advanced exercises in topology are certainly of value. Bending pieces of wire into a series of complex three-dimensional shapes and then sketching their appearance from three viewpoints can teach one a lot.

The four-match problem is a very useful exercise in the inventive use of the physical brain. Match *b* (figure 7.1(b)) is cut to a wafer blade and this is sprung into a split in match *a*, so that the two form a rigid vee. Match *c* is then leant against the apex of the vee so that a pyramid stands up. The problem is to lift the three matches into the air touching them only with a fourth match. The solution is to place one end of the fourth match just below the upper end of match *c*, push the vee *a-b* away slightly so that the upper end of *c* falls between the matches *a* and *b* and then the whole can be lifted by lifting *c*.

Two simple geometrical problems that depend partly on spatial understanding, partly on intellectual freedom are shown in figures 7.1(c) and (d). In figure 7.1(c) one draws the outer square and the half-size one shown by solid lines. First one asks the student to divide the remaining three-quarters of the square into four congruent pieces. He thinks this one out as shown in the dotted lines. Then you ask him to divide the small square into five congruent pieces. The natural tendency is to look for some solution much more subtle than dividing it into five equal strips.

The nine-points problem (figure 7.1(d)) requires one to draw four contiguous straight lines that go through each of the nine points once only. With a well developed visual imagination one should be able to solve it in one's head.

Even more advanced is to construct a wire and string model of Fuller's Tensegrity Mast. This is the simplest shape whereby one can construct a vertical mast of any height solely out of a series of identical rigid elements piled one above the other and joined together only with pieces of cord in

92

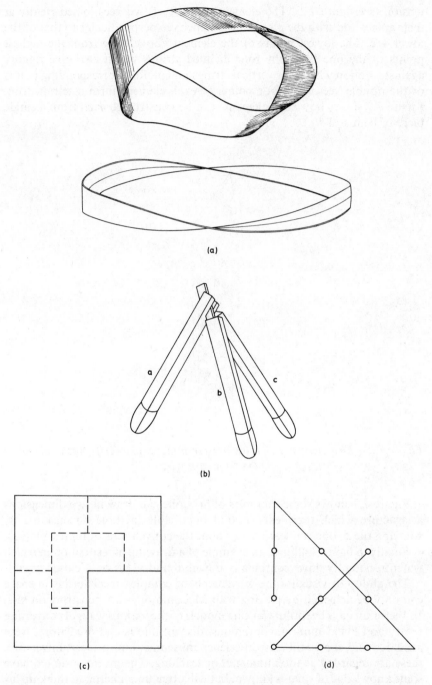

Figure 7.1 (a) *Moebius strip with double cut, single twist (top) and double twist (lower)*; (b) *the four-match problem*; (c) *the square problem*; (d) *the nine-points problem*

tension (see figure 7.2). The elements consist of two vees joined rigidly at their apices and with the plane of the upper vee perpendicular to that of the lower vee. The lowest points of the element above hang from the highest points of the one below by four inclined strings which also give rigidity against rotation when four vertical strings join all four corresponding points of the double vees. The four points of each element form a tetrahedron. Figure 7.2 shows how such elements can be constructed each from a single piece of bent stiff wire.

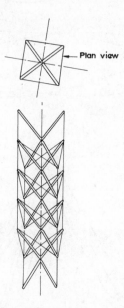

Plan view

Figure 7.2   Five elements of Buckminster Fuller's tensegrity mast; elements bent from stiff wire

Figure 7.3 shows some examples of how one can draw in two dimensions a structure which represents a solid impossible in three dimensions. By studying these, one can learn a lot about the conventions of representation.

Finally in figure 7.4 I give an example of a drawing of a situation in which water apparently flows continually downhill and yet is on a closed circuit.

The ability to visualise the movements of complex mechanisms in space can only be achieved by playing with Meccano or other construction kits. In Victorian days beautiful working models of the various ways of converting rotary and linear motion, continuous, discontinuous and oscillatory, were available and some still exist in science museums, but in most universities these are regarded as too mundane for teaching. The inventor should have some knowledge of what is known, but will often find it better to think up his own special solution to the problem.

Electric power engineers need to develop the ability to visualise the movement of electromagnetic lines of force and mechanical power engineers need

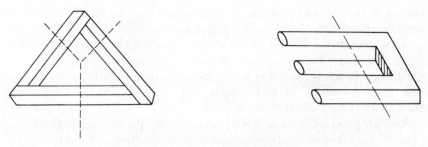

Figure 7.3 Structures impossible in three dimensions; in each case the perspective has different meanings on each side of the dotted line

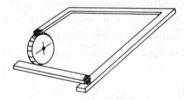

Figure 7.4 Water running downhill in a closed circuit

to visualise the flow of heat, the flow of gases and flame propagation. All these can only be learned by the study of flow visualisation models specially constructed in the laboratory and subject to the correct similarity criteria. Teaching schools at universities and technical colleges who want to encourage their pupils to have inventive ideas in these fields would be well advised to set up two- and three-dimensional water and airflow models in which the shapes of the boundaries can be readily changed and in which the flow is made visible by smoke or tracer particles with powerful planar illumination. Reynolds similarity must be used in such cases with sufficient accuracy to ensure that the degree of turbulence is correct. Two-dimensional models can be used to show flow or force field lines in the two-dimensional solution of Laplace's equation.

THE USE OF TOOLS TO DEVELOP THE ABILITY TO THINK WITH THE HANDS

The use of one-hand and two-hand conventional tools of all kinds for gardening, farming, forestry, carpentry, metalwork, wire bending, wood turning and wood carving provides essential training for 'thinking with the hands'. It gives one a feeling for the shaping and cutting of different materials, the dynamics of impact and momentum, the movement of soft stringy materials, of awkward materials like nails of all sizes, and of leverage and forces in screwing woodscrews. Ratchet screw-drivers and carpenters' braces provide a good feeling for the pressure and problems of reciprocation – the student has to think to realise that the convenience of the ratchet is for a different reason in the two cases; when he understands this, he has connected his sensory knowledge directly with his intellect. Similarly, finding that if you grind a screwdriver with too large a wedge angle between the two driving faces, you cannot hold it in the screw-head slot, leads to understanding why the Posidrive screw system is much more effective than the conventional one – a good example of an invention which would have been first made through the hands rather than the head.

Another good tool problem is to consider why the conventional tinsnips (figure 7.5(a)) tend to bend sheet metal instead of cutting it, if one tries to cut into a sheet at a small angle to the edge. Then the problem is how to redesign it using the same amount of metal so that the cutting blades do not bend apart (figure 7.5(b)). One can learn a lot about handling materials by comparing the use of a pitchfork with a gardening fork to put brambles and garden rubbish on a bonfire or compost heap. Finding the correct weight and shape of mallets, hammers, axes and pickaxes, for purposes ranging from panel pins to breaking concrete gives one a lot of skill in thinking with the hands.

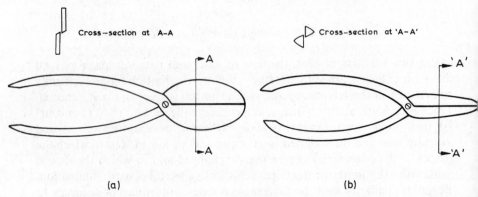

Figure 7.5 (a) *Conventional tinsnips;* (b) *improved tinsnips*

The design of castors, the reason why it is more important to have ball bearings on the vertical axis than on the horizontal one, and the reason for the design of the Shepherd castor, are all problems requiring experiment and sensory observation as well as intellectual analysis. A closely allied problem is why the curvature of the inclined front fork of a bicycle gives straight-line steering stability.

Another instructive field is finding the best object to do a certain job for which there is no special tool. Every handy person will have his own examples. One is the use of a hockey stick to clear a gutter which projected two feet above the top of an 18 ft ladder leaning against a wall.

An extremely valuable exercise is to design and make one's own tools, since this forces one to think about questions of weight, strength and balance and handle convenience that one normally neglects. Some simple examples are

96

(1) A powerful one-hand tool that has proved very useful for many purposes such as digging out and cutting roots, clearing gutters, levering up flagstones and manhole covers, planting, is shown in figure 7.6(a). It is made from 5 mm mild steel with a wooden handle in two halves bolted on.

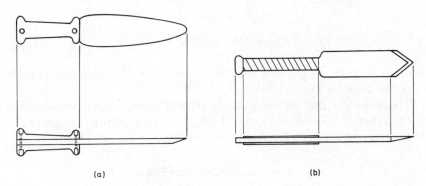

(a)   (b)

Figure 7.6   (a) *One-handed multi-purpose tool;* (b) *a tool for the boot of a car*

(2) A pocket tool with which one can extract plantains, buttercups, dandelions and other weeds as one crosses a lawn. One made by grinding a blunt chisel edge on a 15 mm wide mild steel strip 200 mm long and 3 mm thick and rounding the other end, gives the right degree of strength and ability to cut the roots 3–4 cm below the surface without too much force.

(3) A tool to carry in the boot of the car which does not take up much space and can be used as spade, crowbar or plank to get one out when one driving wheel is sunk in mud or snow. One such tool cut from an old steel gate hinge 10 mm thick, 800 mm long and 75 mm wide has been used successfully on several occasions to help someone in trouble (figure 7.6(b)).

SOME MECHANICAL PROBLEMS AS TRAINING EXERCISES

This section contains a variety of problems which I have come across or had to solve and which are useful for training the ability to think with the hands. All people who are themselves educated in this ability and have the good fortune to be able to put themselves in situations where they have to use this ability (for example in the house, garden, workshop, kitchen or laboratory) will have many examples of their own. However, it may be useful to give these examples to illustrate how discussion and practical examples can help students to acquire this essential skill.

Figure 7.7 illustrates a useful problem to hand round a class. A wide-necked bottle has a thin cork fitting flush in the neck and the problem is to get it out with one's hands. Practically minded people realize quickly that

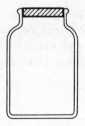

Figure 7.7 Cork in bottle

one has to push it right into the bottle, turn it upside down and pull the cork out between finger and thumb.

The corkscrew, and the gas extractor are well known ways of getting corks out but there is another, the harpoon with a barb which turns at right angles to the stem on a hinge when one tries to withdraw it (figure 7.8).

Figure 7.8 Cork extractor

Some dog collars consist of a length of fine chain with a ring of equal size at each end. Not everyone can see at once how you turn it into a loop which can be slipped round the dog's neck.

A useful practical device is one which can be used to prevent coins from wearing a hole in a man's pocket but allows one to pick out coins with one hand while the other is carrying a briefcase. This consists of a leather false pocket or open purse which fits inside the real pocket so that one's hand goes readily into it and there is plenty of room to feel and select the coins because the front and back are sewn to a piece of chamois leather.

When one has no red tape to tie up separate bundles of documents, one can cut strips across an envelope wide enough to take the documents unfolded and slip one round each bundle. The necessary description can be written on the paper band thus formed, which can be tightened up, if necessary, by folding over the loose edge and clipping with a paperclip.

A friend once put a folding aluminium tube deck-chair on top of some other luggage in the boot of his car and at the end of the journey found that it had slid backwards and was jamming the catch of the boot so that it would not open. By driving out and putting the brakes on hard, he was able to slide it forward and open the boot.

Anyone who has drilled a long thin hole in steel knows how badly the drill wanders out of the true straight line because it is very flexible and has only its stiffness to maintain it. Exactly the same applies if one rotates the work in a lathe and holds the drill in a chuck in the tailstock. So the problem is—how does one drill a rifle barrel which has a very long hole of small diameter and must be exactly true? One answer is to drill a hole in a bar rather thicker than the final barrel and then get a skilled man to straighten the hole by tapping with a hammer as he looks down the barrel. Finally one can turn the outside true with the central hole.

Once I had to make a platform 2 metres above ground which could be moved about on an uneven floor so that I could stand on it to paint the roof beams of an old barn. I wanted it to have four supports on a base wider than the platform and yet not to rock anywhere it was placed. I arrived at the solution almost unintentionally by leaving out the diagonal in the base rectangle (see figure 7.9) so that this was the only one of the six faces of the structure which was not rigid. This enables the two end faces to rotate slightly relative to one another about a horizontal axis through the long axis of the platform; the four feet can adjust to the non-planar character of the floor. The floor thus forms the final element that makes the structure rigid by removing the one remaining flexibility.

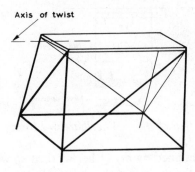

Figure 7.9   Flexible platform stand

In erecting a shed using available materials, I had the problem of fixing the roof trusses single handed. The uprights were railway sleepers projecting 2 metres above the ground and 3 metres apart and each roof truss consisted of two wooden beams 2·5 m long, 10 × 5 cm cross-section. I nailed these beams on the ground using a template and a cross bar to form a rigid triangular frame, but I had to put it into position on the two uprights and hold it while it was nailed in place. Eventually I found that by nailing a beam on the axis of the truss of a length so that it was just off the ground (see figure 7.10), I could lift the truss into place using this as a handle and then prop it upright between two full five-gallon drums. After the truss was firmly nailed to the uprights at the ends, the handle could be removed and used for the next one.

Another country problem arose when the old wheel of a wheelbarrow with 12 mm diameter axle collapsed and I wanted to replace it with a larger wheel with 25 mm diameter axle. I had no machine tools and only a light electric drill capable of drilling 6 mm holes in steel. I fixed oak pieces to the old bearings on the outside with coach bolts through the bearing hole and a woodscrew to prevent each one rotating, and then bent bearings out of 3 mm × 25 mm steel strip to the shape shown in figure 7.11. These were drilled with two holes and fixed to the bottom of the oak pieces with woodscrews. This has worked well for three years since.

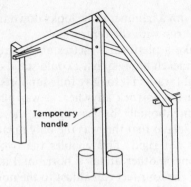

Figure 7.10   Method of erecting roof trusses single handed

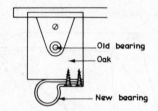

Figure 7.11   Wheelbarrow bearing

An apple-picker designed by my father worked so well that I have since built several of them for my own and others' use. It is shown in figure 7.12. The hinged half circle can be pulled either way by the two strings to grasp an apple so it can be twisted off and falls into the net bag which can hold six or seven apples before one brings it down to empty it.

Figure 7.13 shows a wooden vice capable of grasping round or square logs up to 45 cm diameter for cross-cut sawing. It gives a firmer grasp than a

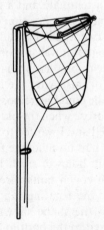

Figure 7.12   Apple-picker

conventional vee shaped sawing horse, because the wedges can be hammered tight to force the parallel jaws on the work, and the pressure of the saw cut works against a vertical surface.

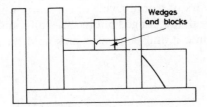

Figure 7.13   *Large wooden vice*

In order to be able to continue my hobby of wood-carving when travelling abroad, I have developed a wooden bench (figure 7.14(a)) which I can strap to my leg, so that when I am sitting down I can clamp the work piece to the bench with a cross bar held by thumbscrews on bolts put through any pair of a series of holes. The bench is made of hardwood 12 × 25 cm with a stop screwed across the end which also has holes for fixing the cross bar so that one can clamp work upright. The work can also be fixed upright in the centre of the bench by woodscrews. Then by placing newspaper on the floor to catch the chips, and unrolling the toolroll on a table beside one, one can do proper woodcarving with a mallet and woodcarving tools. The mallet (figure 7.14(b)) is cut from a flat piece of oak, weighted with a bolt screwed through a hole across the flat end which can also be used as a hammer, and protected with a piece of thick leather nailed on, which also reduces the noise. Another special tool which is very useful in this kit is the combined handle (figure 7.14(c)) for a keyhole wood saw, a broken hacksaw blade and for all the square shanks of bits for a carpenter's brace. A screwdriver

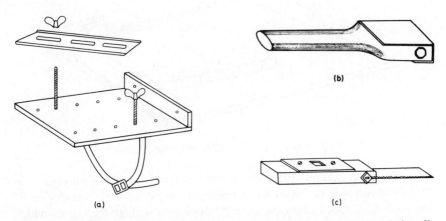

(a)   (b)

                                           (c)

Figure 7.14   (a) *Portable carving bench;* (b) *mallet/hammer;* (c) *handle for brace bits and hacksaw blades*

fitting this handle gives a very powerful leverage and holes up to 25 mm in diameter can be bored in wood with it using flat bits.

A larger tool-kit (figure 7.15) in a teak box, which forms a bench, holds a small machine vice and makes a good sawing horse, enables one to do almost any metalwork or woodcarving or carpentry when on holiday with a car. It has a stout teak top, hinged but recessed all round to fit into the base. A handle is sunk into it and bolts with pins hold the lid firmly closed for carrying.

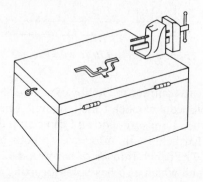

Figure 7.15   Wooden car tool-box

Figure 7.16 shows a portable drawing-board with T-square clamped by a spring clip and a specially designed adjustable protractor which is useful for small working drawings. This chapter is being written on an aluminium clip-board, rounded on the lower left-hand corner so that it fits the body when one writes in a train. The spring clip is placed on the left side so that the paper can be moved up, and one can rest one's writing hand on the board even when one writes the bottom line of the page.

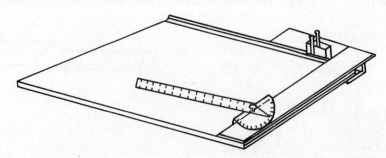

Figure 7.16   Portable drawing board

A very good example for training one's ability to visualise movement in two dimensions is provided by the pin-jointed lattice with one degree of freedom. Everyone knows the simple lattice gate with sliding joints which allows parallel movement in one direction only. The simplest case is shown in figure 7.17(a), and it is easy to visualise the reasons why the movable bar

102

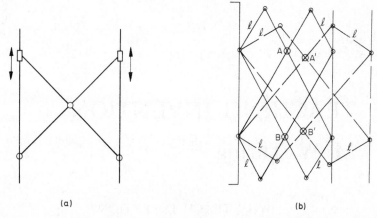

(a)   (b)

Figure 7.17   (a) *Simple lattice with sliding motion;* (b) *simplest pin-jointed lattice with some freedom*

stays parallel to the fixed one and can only move in a direction at right angles to the bars. The problem is to design the simplest mechanism having *only* pin-joints to give the same movement. To solve it one must first realise that the fundamental linking element must be a parallelogram as this is the pin-jointed structure with one degree of freedom. It can be done with two linking parallelograms (linked at A and B) as shown in figure 7.17(b).

By drawing the mechanism in the condition where the distance apart of the two parallel bars is equal to twice the length l of the shorter side of the parallelogram, all the triangles become equilateral and one can see that the long sides must be three times the short sides; also, the distance of the fixed points on a side bar must be $\sqrt{(3)}l$. It emerges that the mechanism consists of a top and bottom half lattice (M and W), each member of one lattice being constrained by the parallelograms to remain parallel to the opposite one in the other lattice. In a recent paper[1] in which this example featured, an additional link is shown joining the centres of the two parallelograms, but it is easy to show that this is redundant as these centres would coincide even if the links A and B were removed so that the right-hand bar could move along its length when the left-hand one was fixed.

REFERENCE

1. Freudenstein, F., *Cardanic Motion*, Fourth World Congress on the Theory of Machines and Mechanisms, University of Newcastle (1975).

8 TEACHING INVENTION

M. W. Thring

WHY TEACH INVENTION?

There are three good reasons why every young person should have their
eyes opened to the methods and joys of inventing.

(1) Inventing is one aspect of the creativeness which is the birthright of
every normal person. Hence every properly educated person should
have had the opportunity to test for himself whether this form of
creativeness is part of his potential, since it is taken as axiomatic
that the true purpose of education should be to enable people to
fulfil all their potentialities.

(2) It is great fun because it is one of the most satisfying forms of exertion
known to man. This is because it requires, like all truly creative work,
the simultaneous use of all three brains.

(3) Some people will use the power of inventing as the central activity
in their work, but all others will find it of considerable use in their
work, their leisure activities, their homemaking and in all new and
unexpected situations in which they find themselves. Invention should
be taught at all ages. The inventiveness of the child just starting to
draw, is usually destroyed by the time the shadow of examinations
looms over them. The worst possible thing a teacher can say to a
child is 'Don't bother your head with new inventions or ideas, it
has all been done already.'

HOW TO TEACH INVENTION

First the teacher must encourage the assumption natural to a young person
of their power to produce original solutions to problems. Then he must help
the pupils to decide on problems worth solving and show them examples
and how to look for problems. At first there will be problems in the home,
the kitchen, the classroom or the car. Later they will find the stirrings of a
social conscience and be concerned with hospitals, cripples, old people
living alone or ambulances and road accidents.

Children have no difficulty in switching off the critical faculty which stifles
ideas before they are born, but by the time they become university students

they have acquired the inhibitions of knowledge and theory with very little understanding of the creative steps by which these were reached. The teacher has therefore to teach them by enabling them to have the experience of inventing, starting with quite simple open-ended problems. It is best done in small groups of less than a dozen, so that every student can give his or her solution to each problem. I have been teaching in this way for eight years and one year two students took out Provisional Patents on ideas of their own by the end of the year.

I give talks based essentially on the material in the first seven chapters of this book, and at the end of each talk we discuss a simple problem requiring thinking with the hands. Any teacher of engineering or workshop skills will have lots of useful examples of his own and probably solutions as good as or better than the ones given here.

Some other examples in this category are

(1) How to drill the hole in a rifle barrel so that it is perfectly straight – a drill of this length and diameter will always wander. The hole has to be straightened by bending the barrel after it has been drilled.

(2) What is the best shape of cutting blade for chopping small branches off trees?

(3) Why does a pitchfork have two tines and a digging fork four?

(4) Why is a shovel at a different angle to the handle as compared with a spade?

(5) Why are hammers of different weight used for different sized nails, and why is a sledgehammer heavier than an axe? Why does a battering ram have to be very heavy?

(6) Design the easiest made substitute for a pair of compasses.

(7) Why is an axe handle shaped as it is?

(8) Why do the teeth on a pocket knife saw face the opposite direction to a normal saw?

(9) A solid steel wedge is ideal for starting a split in a tree trunk but one needs a wider angle wedge for expanding the split, and solid steel would be too heavy. Design a cheap wedge from wood encased in steel strip.

(10) How to design a pair of hand shears capable of cutting a wide sheet of tinplate without the hand getting in the way of the cut metal.

(11) Why is a commercial hand-operated pile driver easier to use than a sledgehammer? This consists of a 5 in diameter heavy steel tube about 24 in long, closed at the upper end by a stout steel disc and with two handles consisting of 1 in steel tubes running the full length parallel to the axis and joined by rounded ends at the top and bottom of the main tube.

(12) What is the leverage of a pair of parallel jaws pliers, and what is the simplest way of measuring it?

I also give them problems to think about from first principles of physics and mechanics, to help them to develop the habit of physical thinking. Some problems in this category are the following

(1) An Eskimo builds an igloo of blocks of ice; the inside surface temperature must be 0°C. How can he undress without getting frozen?

(2) How does the Earth's surface carry the weight of an airplane, a helicopter, and an airship, and over what area is the weight spread?

(3) When one plucks a violin string with the same force first without and then with a sound box, much more sound is heard. Where does the extra energy come from?

(4) Why did a skilled arrow maker test the vibration frequency of each arrow? The arrow had to complete exactly one-half oscillation by the time the feather passed the bow.

(5) Why do people who live in very sunny parts of the world have dark skin when black is the most strongly radiation absorbing surface?

(6) How can one estimate the temperature of an electric cooking oven with a calibrated thermostat?

(7) Is it safer to put the single wheel at the front or the back on a three-wheeled car?

(8) How does hitting the ground with the hands and forearms moving much faster break the shock of a fall on the body?

(9) Why can a man jump so much higher with a pole to help him?

(10) How does a walking stick help one to walk? What is the best length for a walking stick, and why does a simple walking stick with a curved handle give a heavy person with a bad leg nerve pains in the hand?

(11) Why does Bankside power station have four separate flues running up the inside of one large chimney stack?

(12) Why does one see steel chimneys with spiral fins welded up the outside?

Finally each week I give them a longer problem to think about and then they all give their solutions to it the following week. The remainder of this chapter is a selection of such problems with one suggested answer in each case. The teacher is likely to produce better solutions of his own and so may many students. Comments by the teacher of objections to an idea often create a discussion in which several members of the group participate in overcoming these objections by producing further ideas. Several of the problems given below have resulted in such composite solutions in my classes.

THE RUSH-HOUR TRAFFIC PROBLEMS

This hardly needs explaining, but the problem is to suggest feasible mechanisms that would enable the thousands of people who travel daily for distances of the order of ten miles around any big city to do so in reasonable comfort, speed and convenience. Solutions which have been proposed are as follows.

(1) Continuously moving belts with various methods for getting on and off. The earliest suggestion was the series of belts of successively increasing speeds; more recently the Battelle Integrator has been invented. One group of my students proposed a pair of giant discs carrying a flat belt which moved between them so that one side was moved beside the long distance belt at the same speed, and people who had entered the centre of the disc which the belt was leaving could step across at any point between the two discs while those who wanted to get off at this station could step onto the short belt and leave by walking to the centre of the second disc.

(2) Overhead monorail buses with linear motor drive and noiseless magnetic or air cushion suspension.

(3) Various systems for computer-driven minibuses and taxis.

(4) The system discussed in my book *Man, Machines and Tomorrow,*[1] in which each passenger has a luggage trolley cum seat so that no seats need be provided on trains, buses, aircraft and self-drive taxis. Students will usually have a worthwhile discussion on how much commuting can be avoided by improved video telephone, remote printing, etc.

THE LEANING TOWER OF PISA

Again the problem is well known. The tower is built of masonry on a block foundation on ground which is soft for a long way down. It has a tilt and is accelerating as the tilt puts more and more weight on the depressed side. Within 100 years it will fall down unless something is done, but any digging under the foundation is liable to bring it down.

A solution worked out with students is to support it temporarily with a series of air cushions slung between two A-frames so that the air pressure can be raised until the whole lower side is gently supported. Then one can ram 12 in diameter long steel tubes, one at a time, under the foundations with an auger screw extracting the soil as the tube goes in. When a whole platform of these has been pushed under the foundations, concrete can be forced out through holes in the tubes until the foundations are fully supported. Finally the air cushions and frames can be removed.

NON-POLLUTING LOW FUEL CONSUMPTION CAR ENGINE

Battery-operated town cars have been built which essentially operate on power station electricity (which can be generated from coal or hydro power) and give no pollution. However, these have to carry both reagents and thus can never give anything approaching the mileage or light weight of the petrol or diesel engine and fuel. The best immediate solution is the diesel – electric hybrid or diesel – flywheel hybrid. In either of these a diesel engine (preferably with two horizontally opposed cylinders) runs at constant speed and power, and operates a generator. The power of the diesel corresponds to the power requirement for the light car at cruising speed and the diesel is tuned to give very good fuel efficiency, complete combustion and very low noise when running at this optimum condition. When extra power is needed for acceleration it is taken from a storage battery or electrically driven carbon fibre flywheel, which is recharged slowly during cruising or fast when the car is braking or stationary. The rear wheels are driven by two electric motors.

The ultimate car engine, if the car is to survive at all when the world's petroleum is getting really scarce, will have to be a liquid fuel/air breathing fuel cell.

An analysis of some of the large number of possible ways of producing a rotary expansion-type internal combustion engine and their mechanical and combustion characteristics is a very good exercise in the field.

Students may arrive at different conclusions or they may design and invent improvements of these systems, but in any case they have to be given the facts about the weight of air consumed by a car engine, the use of gas and solid fuels, the world's liquid fuel resources and the needs of the two-thirds of the world's population in the less developed countries.

THE AIRSHIP PROBLEM

Airships use under a tenth of the fuel per ton/mile that aeroplanes do, and can be built to carry 500 ton loads. On the other hand 100 mph is about the top speed attainable, they have a ceiling of about 5000 feet and hydrogen is inflammable while helium is scarce. High winds are a great problem when they are tethered, as are differential wind forces and eddies when in flight. One of the biggest problems is landing. If the airship has been running on diesel fuel the weight is considerably less than at the start of the journey and one cannot afford to bleed off helium to lose buoyancy. The weight of cylinders into which the excess helium could be compressed proves to be excessive.

One solution is to provide the airship with an aerofoil wing shape so that it is just heavier than air but can take off by aerodynamic lift at 60 to 80 mph. Another solution is to have hydrogen buoyancy bags above the helium ones and to burn a mixture of hydrogen and diesel fuel in the engines so that the buoyancy can be adjusted by varying the mixture ratio; during flight one would burn a mixture so that buoyancy was just maintained but at the end of the journey only hydrogen would be burnt.

OTHER TRANSPORT PROBLEMS

If we are to continue to have motorways as a principal transport system it is necessary to invent a device whereby vehicles can be allowed to enter the gaps in a stream of vehicles all locked on to a system which moves them at a constant speed and fixed distance apart. This could be operated electro-magnetically or mechanically with a moving rope and hooks; many problems of safety, entering and leaving the system at the required place require inventive solutions.

Similarly it is highly desirable to invent a system whereby the wings and engines and fuel tanks of an aircraft can be detached from the passenger carrying fuselage after it has landed so that the fuselage can form an overhead rail carriage into the city centre, while the power unit is being serviced and sent to the launching ramps.

We need a wheel which can be very hard and rigid when running at high speed on a motorway but can become very soft to give a large road gripping area when one has to apply the brakes. The rim elements should preferably provide both smoothing actions of springs and tyres so that the vehicle has a minimum of unsprung weight.

We also need a device which will enable a car or lorry driver to drive at a reasonable speed in thick fog. A high-power infra-red headlamp and an infra-red image converter screen seems the most obvious solution at present.

WORKSHOP DEVICES

One device requiring inventive solution is the design of a universal vice to grip not only cylinders and flat parallel-sided objects but also spheres, wedges, cones and irregularly shaped objects. Vee-grooves and three-point (or three lines at 120° in a hollow shallow cone) supports on each side, one side being on a central lightly sprung ball joint allowing all three points to grip equally, are quite good for many objects.

Another useful device is the 'third hand', a vice which can be put on a support at any position and any angle to hold a component on a chassis while one works on fixing it. It can also be used to hold a powered hand drill. If all the six degrees of freedom can be clamped or freed by one movement of a locking lever and there is some measure of lifting and overhang counterbalance, the tool becomes much more convenient.

AGRICULTURAL INVENTIONS

With the world's gradual exhaustion of easily won oil resources many of our present devices will have to be replaced by ones that consume much less energy. A light low-power tractor with hind legs to give great traction without panning the soil has already been mentioned. Another defect of present tractors is that they very easily roll over, especially when making a tight turn on a slope, because the centre of gravity is very high in order to give the clearance for the differential on the back axle. A powered system of levelling the body against slopes and centrifugal force would have to work very fast if it was to make the tractor safe. The use of hydraulic motors or direct gearing on the drive wheels would enable the centre of gravity to be lowered as does four-wheel drive with smaller wheels, but these smaller wheels reduce the ability of the tractor to go over rough ground. The Centipede or the twelve-spoked stair-climbing wheels would overcome this weakness and improve considerably on the large tractor wheel or the caterpillar.

With any of these 'legged' devices there is a very interesting inventive problem in the design of the ideal mechanical foot which will slightly dig into mud to provide a good tractive force but splay out to spread the load and shed the mud when lifted out. Probably the mechanical analogue of the cloven hoof is the optimum.

In certain areas it is preferable to punch a very large number of holes 15–30 cm deep in the ground rather than turning the soil over by ploughing. The problem is to design a self-propelled machine which would do this with the spikes entering nearly vertically at the front and being lifted out vertically at the back.

There are many problems in irrigation when electricity for pumping is no longer cheap. Perhaps wind-driven generators, to pump water into slightly elevated reservoirs in the winter when wind and water are plentiful, will be valuable. A vertical axis windmill like the Savonius rotor or one with feathering blades which can be set to feather all round when the wind is too strong will probably need to be invented. Methods of spreading the water on the land when it is needed, to give the equivalent of 2 cm of rainfall all over without the use of high pressure jets, also need developing.

For the hot countries which have monsoon rainfall the problems are to ensure that the water is stored underground or in special covered reservoirs and then pumped out when needed. One way would be to develop a low capital cost solar-operated lifting device, for example on the coffee percolator principle of lifting the ascending column by bubbles of steam.

To irrigate actual deserts the only practicable source of water is salt water pumped up from deep wells or in from the sea. Reverse osmosis and multiple-effect evaporation using fossil fuel are likely to be too expensive, but the use of solar distillation possibly with multiple-effect evaporation can be worked out.

SOLAR ENERGY

At present the use of solar energy is barely economic for domestic water heating, but further inevitable rises in the cost of fossil fuel will shift the balance in favour of such energy. The basic problems in the simple water heating use of solar energy are the high capital cost of the installation and the difficulty of avoiding losing most of the heat from the absorbing surface by reradiation and convection when the water temperature rises to a useful level. The use of selective absorption layers on the water channel, of glass with selective transmission properties and of different ways of circulating the water have all been tried, but are worth further inventive experimental work. The use of a solar heater to give the water one-half to three-quarters of the required temperature rise followed by an electric or solid fuel boiler to top up can correspondingly reduce the electricity requirement.

A honeycomb of thin-walled tubes with surfaces, which are black to infra-red, but white to solar radiation, pointed at the sun can greatly reduce reradiation and convection, but it does not of course increase the rate of energy arriving at the surface which remains somewhat below the free space figure of $1.4 \, kW/m^2$. Mirrors and lens focusing systems can increase this rate a hundredfold or more and can be two-dimensional or one-dimensional focusing. In the latter case the solar daily movement can be in the plane of symmetry so that the focusing system does not have to follow the sun's in travel. Many clever inventions will have to be made if the two-dimensional focusing system is to become practical. In America the 'power tower' concept has been put forward; here a large boiler is located on the top of a 1500-ft tower, surrounded by a large area of plane mirrors which are continuously moved to reflect the sun on to the boiler. An even more advanced idea is a geosynchronous satellite constantly beaming short-wave radio energy to a receiver on the ground. My own view is that solar energy will be used mainly for crop growing in deserts, crop drying by warm air and for small domestic water heating and electricity generation. The electricity generator will probably consist of a battery of solar cells with the solar energy concentrated one-hundredfold or more to save cost in the manufacture of cells of large areas. Solar energy may again be used for cooking in tropical regions using cheap metallic mirrors.

Another very interesting line is the development of chemical reaction chains whereby the solar photon can actually decompose water into H_2 and O_2.

MAN-POWERED FLIGHT

It is certain that conventional attempts to solve this problem by a combination of bicycle, pedal-operated airscrew and glider-like wings, are doomed to failure. However the different approach of a pedal-operated device which pulls air backwards along the upper surface of heavily cambered wings may well succeed. This would give lift even when the plane had no ground speed since lift arises from the Bernoulli suction due to the airspeed on the top of the wing. Thus the athlete would not have to reach a high bicycling speed against the air resistance before taking off. One way of doing this is to use a biplane with a very light belt running round the upper surface of the two wings – the belt carrying a large number of short propeller blades at right angles to it.

DOMESTIC INVENTIONS NEEDED

(1) The domestic equivalent of the flexible grinding shaft which can be used to scour, scrub and polish with a device to reduce rotary or reciprocating reaction (for example, two discs in opposed rotation or balanced twin reciprocators), and a device to feed soap or polish without throwing it around.

(2) The automatic kitchen cupboard that brings the required shelf to reach when a button is pressed. Buckminster Fuller exhibited one forty years ago based on two endless belts around two pulleys on the floor and two on the ceiling. The problems are the speed of response and the rather deep bulky cupboard required unless the design is very good.

(3) A pan scourer fixed to the sink so that burnt-on food can be scoured away rapidly.

(4) A small domestic continuous dishwasher in which one sets the dishes or saucepans on a tray on one side at the top and starts the machine. As a succession of trays go down one side they are rinsed clear of food residues and washed; on the other side they are dried and come out at the top in a pile of trays for unloading.

Teachers and students will readily be able to contribute their own problems and solutions in this field.

SURGEONS' TOOLS

Stapling machines using molybdenum staples to replace surgical sewing already exist, but there is a real need for a sewing machine which will rapidly put in stitches of flexible thread exactly where the surgeon needs them. An instrument like a pair of pliers which can be preloaded and can then insert several stitches when it is placed in position and operated can be designed. The problems are passing the thread through in the right way (for example by a front-eyed half circular needle) and fastening the stitch (for example sealing the two ends together by heating a suitable plastic).

Powered chisels and saws for surgeons also offer scope for inventive improvements.

Many ingenious micro-manipulators have been developed which can be used with low-power binocular microscopes for simple operations in dissection, surgery and even industrial microcircuit construction. We constructed a special table with adjustable hand-rests for micro-surgery and a foot-controlled micro-scissors and forceps so that the hand is not shaken by having to operate them.

However, the student of invention can learn a lot from designing his own micro-manipulator to give the necessary seven controls for each tool (three space dimensions, three space orientations, one tool operation) all scaled down by say 4–1.

SCEPTROLOGY

Sceptrology is the science and technology of crutches and contains numerous important inventive problems.

(1) A battery-operated device to raise a collapsed wheelchair to the roof of a car and stow it there so that a cripple with strong arms can enter, drive and leave a car without aid.

112

(2) A transfer device from electrically powered wheelchair to car seat.

(3) A device to help people with arthritic hips to get on and off the lavatory by themselves.

(4) A similar device for the bath. (We have made one with a hydraulically lifted seat operated by a lever.)

(5) A device so that a person with one weak hand can open tins, not using electric power. Also a safe kettle and teapot pouring device.

(6) An extended hand with sufficient movements and feel, operated by the same movements of the real hand, to pick things up from the ground or from a high shelf.

(7) A device to give outside warning when an old person living alone is in trouble. The most ingenious solution I heard of was a system which sounded an alarm if the lavatory cistern was not flushed for twenty four hours (but even this may be too late).

(8) A device for lifting a heavy inert patient from the ground. Don Jordan of Australia has developed a tubular frame in two halves which can be assembled around the patient and with enough plastic strips pushed under to take the weight.

(9) A device to help blind people to walk about. Walking sticks which give an audible note in a pair of earphones, the pitch of which varies with the distance away of the solid object at which the stick is pointing is one solution which has been developed. A device so that a blind person knows when a teacup is full – conversion of a pocket watch or small clock so that a blind person can tell the time. Writing device that produces a line that can be felt.

(10) Device that can signal printed letters to a blind person, one at a time as he moves it along a line of print.

(11) A page turner.

POLLUTION PROBLEMS

(1) A dry cleaning device for taking sulphur dioxide out of hot flue gases – dolomite chips can be used but it is very expensive to dispose of the resulting calcium sulphate and magnesium sulphate double salt.

(2) A practical device for removing all the impurities from river water before it flows to the sea and then separating them into recoverable heavy metals, fertilisers (phosphorus, potassium and humus) and other useful chemicals.

(3) A thermal precipitator combined with a regenerative heat exchanger for glass tanks and steel melting furnaces.

(4) A device for turning city night soil or farmyard dung into dry, clean fertilizer powder, free of lead.

PROBLEMS CONNECTED WITH ACCIDENTAL FIRES

(1) When a fireman directs a water jet into a burning room he has to get quite close because the water jet breaks up in quite a short distance

into a spray which spreads widely. One solution is to put a gelling agent into the water so that it forms a solid rod in flight like the wartime crocodile flame thrower (a good example of a wartime problem being solved because sufficient money and effort was put into it). Then a jet could be thrown through a tenth floor window from the ground. Another alternative is to pack a liquid extinguishing agent into plastic sausages which could be fired by a CO_2 gun – the plastic would burn off in the fire and release the extinguisher.

(2) Fires in large unoccupied factories or warehouses could be detected and extinguished by a robot night watchman which travelled round the building on a predetermined path or followed a track painted on the ground or travelled on an overhead rail. This could carry two expensive fire detecting instruments, for example it could scan by infra red and then check by flame ion detection. A jet of fire extinguisher could be directed at the fire and an alarm sounded.

(3) A telechiric fireman, water-cooled and with trailing hose and communication cable, could go into the heart of a fire and climb stairs and extinguish the fire at its source. It could also rescue unconscious people and put them inside a safe miniature cabin to bring them out.

(4) Rescue apparatus for saving lives in skyscraper fires.

SAMPLING GEOLOGICAL DOMES

It is fairly easy to determine the presence of a dome of impervious rock underground by the reflection of waves from an explosion, but in order to find out whether the permeable rock contains oil, gas or only water it is necessary to drill a test well. If one could find a cheaper and quicker method of sampling, the whole system of prospecting for new oil would become much cheaper.

One solution would be to develop a drill powered at the nose – perhaps a percussive drill with a free-piston explosion engine running on liquid fuel and liquid oxygen pumped down to it. Such a drill could drill a hole of much smaller diameter. By lining the hole with a steel strip overlapping helix, drawn in from the bottom, it would not be necessary to withdraw all the drill tubes every time one has to change the bit.

ROBOT PROBLEMS

The development of robots to relieve humans of excessively tedious tasks requires the inventive solution of a number of problems.

(1) The design of a mechanical hand with only one controlled movement which can do nearly all the handling jobs that the human hand can do with four nearly independent movements for each digit.

(2) An optical method of locating a flat object of predetermined size and shape on a table and observing its orientation. The 'sledgehammer'

method is to use a TV camera and a large computer. Methods using 1–4 photoelectric cells and mirror vibration scanning, a zoom lens and masks are much more practical. Such a system can be used for robot palletising of components.

(3) The design of a robot that can be instructed by a central computer to go to any bin in a three-dimensional assembly of bins and take a number of packets from the bin.

(4) The design of an inspection robot to reject faulty components by comparing them with a template.

(5) The design of mechanical muscles for the operation of robot limbs.

MISCELLANEOUS PROBLEMS

(1) A telechiric machine to drill oil wells at the bottom of several hundred feet of sea water by remote control from the shore. It would have to do all the operations at present done by men and machines on the drilling platform.

(2) A 500 kW high-efficiency electricity generator to provide power in local areas of towns or villages with only daily maintenance and no full-time operator. If coal was the fuel this could be converted into a gas, such as blue-water gas at a central converter and piped up to a distance of fifty miles to the generators.

(3) A telechiric machine for rapid rebricking of electric arc furnaces capable of installing half-a-ton arrays of bricks at a time.

(4) A loom not requiring a mechanical shuttle. One has already been developed using an air jet from a nozzle to blow the thread through.

PROBLEMS FOR LESS DEVELOPED COUNTRIES

Schumacher (*Small is Beautiful*) has pointed out that the less developed countries are short of factories, production machinery and fuels, but not of manpower. They need to have machines that will enable them to be more self supporting at a considerably higher standard of living. He has called the technology that needs to be developed 'intermediate technology', and it contains many fascinating and worthwhile problems, including the following.

(1) Solar operated pump.

(2) Vertical axis windmill for pumping. This does not have to be turned into the wind and can be very simply mounted on a tube with guy-ropes, the power coming down a central shaft which is an extension of it. This can be done with fixed aerofoil or flapping blades.

(3) Irrigating deserts by solar distillation of salt water condensing on the inside of transparent Polythene roofs.

(4) Water preservation in the tropics where there is a short rainy season. It is necessary to avoid excessive evaporation losses such as would

occur if one used small open reservoirs all along the river course (for example by using a layer of small floating particles). It is also necessary to avoid the evil consequences of a big dam at the head of the river (prevention of valuable silt carry over, filling of the reservoir by silt, disease due to upsetting good cleansing).

(5) A power system that can be run on any local combustible, such as wood branches or non-compostible agricultural refuse. For example, a simple cast-iron steam engine equivalent to a car engine with boiler, or a charcoal operated gas producer to operate a spark ignition engine.

(6) A 2 hp mechanical horse with two hind legs instead of wheels to give high traction (for example, for rice paddy fields).

REFERENCE

1. Thring, M. W., *Man, Machines and Tomorrow*, Routledge (1973)

9 DEVELOPING AND PATENTING AN INVENTION

Part 1, M. W. Thring

THE FOUR STEPS OF DEVELOPMENT

It is certainly true that more than ten people invent a new process or product for every one who brings it to fruition. Every inventor who has been associated with a new development knows that after he has correctly formulated a problem and invented a really new solution to it, his work has only begun. The next stage is for him to convince at least a few people in positions of power that his invention can actually work. After that he has to demonstrate that the invention can work economically in comparison with existing processes or products. Only then can he proceed to the final stage of putting it on the market and finding out whether it does in practice satisfy the human need concerned.

Actually, as illustrated in figure 9.1, there are four stages of development, but in many cases the first or lowest stage has already been done, or is not necessary, or is done long after the invention has been successfully developed. This lowest stage is the laboratory bench work needed to supply the fundamental physical or chemical coefficients for feasibility studies. Thus for example a new idea must be tested at this stage as to whether it is contrary to the accepted interpretation of the first and second laws of thermodynamics, that is produces energy from nowhere or from heat at ambient temperature. If it is a chemical reaction one needs to know the sign and magnitude of the reaction energy (whether it is endothermic or exothermic) and the activation energy. Two examples of inventions in use long before these were known were the use of fire (the combustion reaction) and the manufacture of glass. The physical and chemical structure of glass has only been determined in the laboratory in the last fifty years, whereas the Egyptian Pharaohs had glass ornaments made by fusing sand and potash and salt. An example at the other extreme is provided by the production of electricity from controlled nuclear fission. The atomic bomb could be made from the laboratory discovery of uranium fission which produced neutrons enabling an explosive chain reaction to occur. But control of the process for peaceful uses required a detailed knowledge of how the two isotopes U^{238} and U^{235} behaved in fast and slow neutron bombardment.

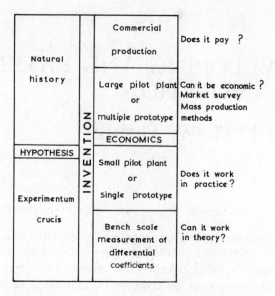

Natural history	INVENTION	Commercial production	Does it pay ?
		Large pilot plant or multiple prototype	Can it be economic ? Market survey Mass production methods
		ECONOMICS	
HYPOTHESIS		Small pilot plant or Single prototype	Does it work in practice ?
Experimentum crucis		Bench scale measurement of differential coefficients	Can it work in theory?

Figure 9.1   The four stages of development

The whole transistor and semiconductor industry had to be based on a series of fundamental discoveries on solid state physics obtained in the laboratory, just as earlier the thermionic valve had been based on Edison's observation of negative electricity flowing from a hot filament.

The second rung on the ladder of development can never be avoided. This is the stage where the invention is first realised in hardware. It passes from conception to reality when one makes a working model and one has to answer the question 'Does it work?' In the case of a mechanism such as the centipede walking machine no new fundamental knowledge is needed and one can proceed straight to this stage.

This first working model or single prototype does not have to be of the same size as the final intended unit and is often considerably smaller to reduce costs and speed up construction and alteration. In some cases it may be conveniently larger than the final unit. In either case one has to be sure that the successful operation of the scaled prototype is a good indication of the successful operation of the final device. In the case of MHD or nuclear fusion where the problems are on the walls and the performance is in the volume, scale up makes the operation more successful, whereas in a case such as caterpillar tracks for mud crawling the tracks occupy an increasing fraction of the ground area covered by the body and so there is an upper limit to the weight. Similarly one can develop a fuel cell for a few kilowatts but it is very difficult to envisage a 1000 MW fuel cell assembly. Thus the fuel cell could be used to generate electricity in each house but not in a huge central power station; on the other hand MHD or fusion will require power stations even bigger than our present ones. A knowledge of the laws of scaling[1] is therefore very important to the engineer who intends to develop an invention.

This stage is more appropriately called the small pilot plant where one is developing a new chemical process, a new energy conversion process or a new chemical product. For example, when I was trying to develop a continuous steel-refining process in Britain, the small pilot plant was the smallest flame-heated channel which could operate continuously without being quenched by heat losses: we designed this at 3 ton/h with a channel only 10 cm wide. Again the basic aim is to demonstrate that the process can be made to work in real space and at a real rate of throughput.

Until this work is complete it is nonsense to make any economic assessment because one can have no idea of the capital cost for a given output, the materials that must be used or any other of the questions such as reliability and maintenance which must be answered before the accountants can do sensible calculations. This may be called 'the fallacy of premature economic assessment'. A classical example came when I submitted a proposal for a mole miner in 1965 and the Coal Board said it would be uneconomic. It was not until 1970 that I learnt that they based this conclusion on the fact that when they cut a roadway through a coal seam by man-operated machines and prop it with man-handled arches the value of the coal obtained does not pay for the cutting of the roadway. Since no working prototypes of my remote-controlled machine had been built this assessment was clearly irrelevant.

It is certainly true that Parsons's steam turbine, Edison's electric lamp filament, the steel-making processes of Bessemer and Siemens and almost all the other great inventions would never have been successfully developed if they had been subjected to the fallacy of premature economic assessment by an official committee.

Thus we can say that economic questions begin to be answered only when we reach the large pilot plant or multiple prototype stage. These economic questions relate to

(1) The cost of manufacture, for example what are the cheapest materials that will give reliable results; what is the cheapest method of forming the parts for the numbers expected to be produced (casting, machining, welding or hot forming)?

(2) The costs of operation – expected efficiency and fuel consumption, throughput and other performance factors, maintenance requirements and likely outage time.

(3) The likely lifetime – very difficult to estimate because it depends on unpredictable factors in the future economic situation such as inflation and depression. However the development engineer should be concerned to make all the parts last as long as he guesses the invention is likely to be wanted.

If any object involving a new invention is going to be mass produced it has been found by bitter experience that it is extremely costly and time-consuming to try to omit the multiple prototype stage and go straight from the single prototype to mass production. It is this mistake which often causes the car makers to have to make an alteration to thousands of cars already sold because these have developed a fault which was not shown up on the single prototype. The multiple prototype must involve ten or more models made with the same materials and, as far as possible, by the same methods as those planned for the final model. The purpose is not only to give a reasonable chance of spotting weaknesses but also to give the designers an opportunity to have second thoughts on the materials and methods of construction before it is too late. Since weaknesses in vehicles or other systems subjected to millions of load cycles are usually due to fatigue or stress corrosion which show a small proportion of failures long before the average, it is necessary to operate these multiple prototypes for many months or use a very large number if one is to avoid a significant early failure rate in practice.

PROCEEDING THROUGH THE STAGES

Figure 9.2 illustrates the broad relationship in time of the four stages of development. Zero time is the moment when the inventor obtains financial agreement to proceed with the development of his invention. Any bench-scale research which is essential should be started at once and so should the detailed design of the single prototype or small pilot plant. The small pilot plant is essential in the case of a new process, for example in chemical engineering, agricultural and food engineering, steel making, or fuel utilisa-

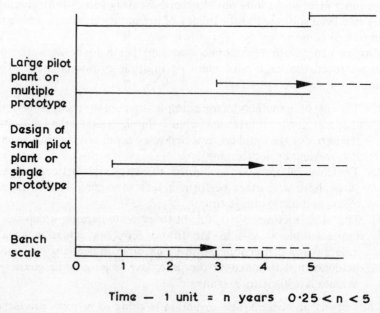

Figure 9.2   Development in time

120

tion. In the cases of the invention of a new machine then the single prototype is the second stage.

Only when the single prototype has demonstrated that the invention is feasible, or the small pilot plant has given the first performance figures, can the third stage be designed using this information to guide the designers. Since there will be still a considerable use for the second stage in trying out ideas for major modifications the second and third stages will show a considerable overlap. The time scale of figure 9.2 is subject to a multiplying factor. In a very active industry which can do development experiments quickly such as the electronics and computer industries this time is very short. Usually it is about the transition point between the second and third stages that the inventor ceases to be the best person to be in active executive charge of the work. It is very rare for the inventor to combine in himself, with equal success, the three roles of true invention, finance raising and detailed commercial development. Thus it is much more common for the invention to require two 'godfathers', one to provide the necessary financial and business acumen and the other to have the patience, persistence and human handling ability to carry the idea through to final commercial success.

Development of any really major invention is necessarily a gamble at every stage but since each stage costs at least ten times as much as the one before it is necessary to have some criterion to decide whether one should start the next stage and whether one should continue one stage for another year when the costs rise unexpectedly – as they always do in development work. A rough guide is the following formula:[2]

$$\text{if } \frac{A \times B}{C} \geqslant 3, \text{ then proceed}$$

which gives at least qualitative expression to all the three important factors in the decision. These are

A = Expected benefit expressed in money if the final invention is successful

B = Probability of success – a simple fraction

C = Expected cost of the remainder of the development

Of course the assessment of all three of these factors is largely subjective and must be carried out by a kind of judicial or committee procedure, but it is very important that this should neither be biased in favour of, nor (as is much more common) against the new invention. It has been said by Fielden that the Pilkington float glass process to make fire-polished sheet glass by casting it on molten lead would never have been achieved if the inventor had not been a member of the Pilkington family, since the development ran into such costly snags that the pessimists would certainly have stopped it before the crucial problems were solved. In chapter 2 we have seen many examples of the attitude of the establishment and even of elderly inventors to new ideas.

In assessing factor A the expected benefit should ideally be proposed by the financial 'godfather' who must have enthusiasm as well as business sense. Ideally he would be the development director of the firm which is either already doing the work or hopes to take it over at the multiple prototype or large pilot plant scale. If he also serves as judge or chairman of the development committee he can ensure that the 'prosecution'–all the members of the committee who prefer to do things as they have always done them or are against all change on principle–are not allowed to be unduly pessimistic and block the work. In assessing factor B the inventor must be the 'counsel for the defence' but his assessment of the probability will always be much too high and that of the 'prosecution' so low as often to be zero. Here the chairman has a vital role in sorting out the objectivity of the arguments on each side and allowing the inventor to have time to make the subsidiary inventions necessary to carry the invention past the difficulties that will be so strongly pressed by the 'prosecution'.

Factor C should be a straightforward estimate by the project engineer who is preferably not the inventor but is the second 'godfather'. He should ideally be in the position that he will get rapid promotion with the development if it succeeds, but will share the burden of failure with the whole development committee if it fails. If it is stopped for political reasons or a financial change completely outside the control of the project engineer, such as the unexpected drying up of a source of supply of raw materials, this should not count as the slightest black mark in his career.

The reason for requiring that the dimensionless ratio $A \times B/C$ should be greater than 3 and not 1 is to allow for uncertainty in the assessments and for the fact that the successful inventions have to subsidise the unsuccessful ones. Since the cost of stage 1 will be of the order of hundreds of thousands of pounds, even in the cases where it has to be done for a big invention, there is no great difficulty in justifying stage 1 which is mostly and very properly done in the Engineering Departments of Universities. It is often financed by industry or Government without any expected financial benefit at all, or for a largely undefined possibility such as the laboratory work on leaf fractionation which has been done for nearly fifty years and is only now becoming economic.

STAGE 2

The start of stage 2 (the single prototype design, construction, testing and modification) is the point at which one must set up at least the rudiments of a development committee and make the first rough application of the procedural formula. The cost of this stage may be as low as a few hundreds of pounds, or even less where the invention is such that the inventor can construct the first series of prototypes in his own workshop. This model should be made of the softest and most easily shaped materials having just enough strength to demonstrate the working of the main idea without worrying about subsidiary effects like vibration or fatigue. These effects need to be studied later before the invention is shown to anyone except the engineer godfather.

Other projects in the mechanical field, such as powered steering for cars or Ferguson's powered tool holder for farm tractors, require to be made, even

122

for the first working prototype, with proper engineering drawings, stress calculations and good machine-shop construction. Even here however a mock-up which can illustrate the mode of operation, made from Meccano or one of the modern prototype kits, is very valuable. I often make a non-working mock-up out of sheet aluminium, wood or wire just to have something in three dimensions to discuss with my draftsman. With a very advanced idea, such as the remote mechanic for coal mining, a small non-working model can be a great help in trying to find a financial godfather for the idea.

Projects in the chemical engineering, power conversion or process metallurgy fields always require a small pilot plant, which will operate with suitable materials of construction for a series of short trials with the actual raw materials at the correct temperatures and pressures. This work may take several years and cost hundreds of thousands of pounds.

The type of experiment required to be carried out at stage 2 (the development experiment) is fundamentally different from either the classical single variable experiment or the statistical multivariable factorial experiment. In the classical experiment one independent variable is changed, usually over a whole series of values, and a curve is plotted for the dependent variable while all other variables are held constant. One has to design the experiment to look for hysteresis effects (effect of previous value of independent variable), time effects and random scatter.

In the factorial experiment one usually varies several independent variables at two levels, each in a planned series of experiments, so that as well as the primary effects one can detect first and even second order interactions.

In the development experiment the procedure is to run one trial with all the independent operating variables at the settings corresponding to the expected values in the final system and with all the design variables made to what is considered to be the realisation of the invention most likely to work. After the trial a small working party, which must contain the inventor and the project engineer, holds a post mortem and decides how to change as many of the design variables as they consider are likely to improve the performance. They have to bear in mind that sufficient success must be obtained after a series of experiments, very limited by both time and money, to enable the project to be extended to stage three when the development formula is applied.

It is largely dependent on the extent to which the engineers (especially the inventor and the project engineer) produce subsidiary inventions to solve the problems and overcome the failures during this series of development experiments, whether or not the invention proceeds to stage 3. At stage 2 the odds are of the order of $B = 1/10$ but since C is still small this does not matter. By stage 3 they should have improved to $B = \frac{1}{3} - \frac{1}{2}$.

REFERENCES

1. Johnstone, R. E. and Thring, M. W., *Models, Pilot Plants and Scale-up Methods in Chemical Engineering*, McGraw-Hill, New York (1957)
2. Thring, M. W., 'A Workshop for Inventions', *New Scientist* (1970)

Part 2, E. R. Laithwaite

INVENTIONS, PATENTS AND PROCEDURES

Patenting ideas is one of the quickest and easiest ways of losing money! It seemed to us that one of our duties to you, the reader, was to give you the benefit of our knowledge and experiences in a book on invention. An invention and a patent are essentially different. The first is based on ideas and imagination. It has been said that inventing is the one thing that separates men from other mammals. (We would both have additional ideas on this thought which have no place here!) The second is a facet of common law.

The first thing to realise is that millions of patents already exist. Couple this with the thousands of millions of humans who have had opportunity to do new things, and the chance of having a unique thought appears no more than that of being an outright winner of a £300000 football pool; but if you never fill in a football pool you are *certain* never to win! It is surprising how simple are some of the successful inventions of only last year.

If we can put the emphasis in its proper place however, then we must advise you to err on the side of losing an opportunity rather than on losing your life's savings, so take into your confidence someone with experience in the subject, before you spend a lot of money on an idea. To this end the Provisional Specification was developed so that on payment of £1 you can obtain a year's protection from 'pirates' and businessmen of little scruple, and yet, simple though it sounds, the £1 does not provide you with someone to set out your invention in a way that restricts your rivals but not yourself, and patent writing is rather like walking through a minefield. The qualifications of a registered patent agent are very demanding, necessitating a first degree standard in at least one specialised technology and a knowledge of the law equal to that of a barrister. Naturally therefore, an agent's fees are not trivial.

We thought that a way of introducing a section on patent law that might be both interesting and instructive would be to write a 'potted' history of the subject in the next few pages. This will introduce the beginner to the terminology in a natural way and will, we hope, still be of interest to those with experiences of their own.

ORIGINS AND CONCEPTS

The origin of patents was, like the origin of invention, the origin of man, the origin of the Earth and the origin of the Universe, obscure. Although an entirely human concept it did not begin with a fanfare of trumpets or a single act of parliament. Goethe insisted that the invention patent 'originated among the English who draw profit and advantage out of everything'.

Although Benjamin of Tudela mentions an exclusive privilege for dyeing cloth in the twelfth century, most authorities are agreed that the origin of patents (if they are to have one) lies with the mayor of Bordeaux who granted a privilege in 1236 giving exclusive right for fifteen years to make cloth following Flemish or English methods. (The period of fifteen years is almost relevant to modern practice but only by coincidence.) The fact that this privilege was confirmed by King Henry III (who was also Duke of Gascony) probably strengthened the idea. Frumkin points out that this was a patent-like grant, but not a patent in the modern sense, since no specific invention was involved.

Edward III then began giving 'letters of protection' to Flemish weavers to attract them to England, the first being in 1331. The first English 'patent of invention' is said to have been granted in 1449 by Henry VI to John Utynam who made stained-glass windows. This grant corresponds to what is now called an 'importation patent' now only used in Belgium, Spain and South America.

The real beginning of the English patent system is usually said to have rested in a patent for Normandy glass to Henry Smyth in 1552, although E. W. Hulme, in his authoritative papers on the history of patents, has as his 'No. 1' a patent granted to two immigrants in 1561 for making Castile soap. The first seven patents listed by Hulme were all granted for 10 or 20 years. Continental patents were traditionally for 7, 14 or 21 years. Later the multiple of seven became universal. Printed records and indexes of English patents of invention from the year 1617 were compiled in 1853, but as Gomme observes in his excellent survey of the origin and growth of the system,[1] 'These are, however, to modern eyes, but the bare bones, giving us the facts of the grant but very little about the actual nature of the invention and its method of working. The essential document for this purpose is the patent specification, which, barely mentioned hitherto, must now occupy our attention. Coming late on the scene, it is today the most widely known and important of all patent literature and, in itself, a document of great technical and commercial significance.

'The patent specification can be defined as a document containing a description of a patented invention sufficiently full and detailed to enable the invention to be understood in its nature and applied in practice by persons skilled in the particular art without further experiment, separate from the Letters Patent themselves but subject to a clause in them, the non-compliance with the terms of which would serve to invalidate the grant. The specification did not become a regular feature of English patent practice until the first half of the eighteenth century, but from this time to the year 1852, and in certain circumstances to the year 1883, the validity of a patent was made dependent on the lodging within a stated time of full specification, by the insertion of a clause in the grant . . . .'

The clause referred to is to the effect that the nature of the invention and its method of working must be inrolled in the High Court of Chancery or the Letters Patent would be void. The first patent specification is said to be that of John Nasmith in October 1711 for the preparation and fermenting of wash from sugar and molasses. The modern interpretation of the specification is the insertion in the grant of a proviso that requires the description to be delivered within a specified time after the sealing of the patent. This practice lasted for over one hundred years until 1884, but the Patent Law Amendment Act of 1852 made a radical change that 'evolved' rather than was deliberately directed towards modern practice.

From 1852 the rights of a patent were dated from the date of *application* instead of from the date of 'sealing' (acceptance). Inventors were then permitted to lodge 'complete' specifications after the initial application, safe in the knowledge that their idea was either 'no good' or 'protected from birth', so to speak. The old idea of a specification produced within a set time after sealing was retained as one alternative but with the condition that a 'Provisional' (shortened form of) specification had to be submitted *with the application*. The second alternative was the old method of submitting a 'Complete' at the time of application, but what amounted to a penalty clause was then inserted in that the application was void if the application 'did not fully disclose the invention'.

Together these two changes causes a drift towards the modern standard practice of a 'Provisional' followed by a 'Complete' after twelve months. Like patents as a whole, the 'Provisional Specification' had no sudden origin. Provisional Specifications that were not followed by a complete specification were printed up to 1883, but have not been published or accessible to the public since then. The provisional specification enables the applicant to stake his claim to priority for an invention which is to be described in more detail in the complete specification at the very small fee of £1.00. He secures his date and is given twelve months (extendable to fifteen months on payment of a special fee) to make technical and commercial enquiries so that he can develop the idea fully before he files his complete specification. But, he must *not* include new items in the complete that were not disclosed in the provisional. He can however apply for a 'patent of addition' if a subsequent invention relies very heavily on the one already filed.

PROPOSED REVISIONS 1975

At the time of going to press a Government White Paper on 'Patent Law Reform' [Cmnd. 6000] proposes certain changes, implementing the recommendations of the Banks committee (July 1970) and enabling the U.K. to ratify the Council of Europe Convention on the unification of certain points of substantive patent law, the European Patent (Strasbourg) Convention and Patent Convention Treaty. If these changes become law, the most important of them will be that

(1) Every specification must be the subject of a separate application (at present, the 'Complete' needs no further application).

(2) An application may proceed in its own right to grant and/or serve as a basis for priority in respect of a later application and specification filed not more than twelve months later.

(3) Any application may claim priority from one or more earlier British applications and/or one or more European applications of foreign applications filed within the preceding twelve.months.

(4) It will be possible to file an application, accompanied by a specification for a nominal sum (such as is now done with a 'Provisional'). The application need not include claims (as is now the case with a 'Provisional'), but within twelve months of filing or of application the applicant *must* request a search and pay the fee (which is large), and this is basically little different from the present 'Complete' for a set of claims must be sent if not sent with the 'Provisional'.

(5) Alternatively, he may, within twelve months, file a fresh application, claiming priority from the earlier application (this could ease the situation regarding the addition of new ideas). A further nominal fee is required and the earlier application would be abandoned on request.

(6) The search for novelty and obviousness will be separated from the technical examination, to speed up the process of rejection.

The Patent Comptroller will still have power to require amendment before grant and to reject if not amended to his satisfaction, as at present. Other changes are more specialised and the inventor who wishes to write his own specification (as opposed to paying an agent to do it for him) will need to read the White Paper and supplementary booklet.

It is hoped that from this 'potted' history, many of the ideas incorporated will have become implanted in the mind of the reader, so that the following paragraphs (which only relate to pre-1975 practice) need not be spelt out in such legal or formal terms as would otherwise have been necessary.

CLAIMS

Whereas a provisional specification simply *describes* an invention in order to establish a 'priority' date, the complete specification ends with a list of claims, beginning with the phrase 'What I/we claim is . . . . There follows a series of numbered claims beginning with the widest possible view of the invention and ending usually with a claim so particular to a specific device that if it is the only claim allowed it is of no commercial value and could hardly even be called a Registered Design. It is usually phrased 'Any apparatus constructed according to the accompanying drawings.'

Claim number 2 may be dependent on claim number 1. If so, it begins with words such as 'A motor according to claim 1 in which . . .'. On the other hand, claim 2 can be a different *embodiment* of the same idea and completely

127

independent of claim 1 (perhaps a tubular version of a machine which in claim 1 was a *flat* machine). Having narrowed the extent of the first claim in a subsequent one (claim 7, for example) there may then follow claims 9, 10, 11, each of which begins: 'A mechanism as described in claim 7 . . . .' Claim 8, however like claim 7 may have been a special case of claim 1, and therefore independent of claim 7. When the list of claims dependent on 7 is finished, there may follow another list of claims, for example, 14, 15, 16 . . . , each of which begins 'A mechanism as described in claim 8 . . . ' . The system is infinitely extendible.

It can also be seen to be a form of classification not unlike the biologists' dichotomous tables or even nearer to a genealogical table. The only complication arises where, for example, claim 24 begins: 'A mechanism as described in claims 1, 2, 8, and 17 wherein' or even: 'A mechanism as described in any of the foregoing claims wherein . . . .'

Last of all come the most detailed claims all of which are almost always independent and which read: 'A device constructed and arranged substantially as herein described with reference to and shown in figures 2, 3 and 5 of the accompanying drawing', and further similar claims until all the figures have been referred to. Alternatively, the last claim can read simply: 'A device constructed and arranged substantially as herein described with reference to the accompanying drawings.' The reason for splitting the figures into specific groups is that some of the figures may not be granted as novel arrangements and therefore the deletion of some may not invalidate the whole. The patent writer has to judge for himself whether or not the figures will so 'group'.

The whole idea of drafting claims follows very closely other legal practices, where it is not so much a question of fact, as of what you think you can get away with! To take a simple and exaggerated example to illustrate the point, suppose a man wishes to invent a new kind of door hinge that automatically closes the door. His submission claims may read

(1) A door hinge wherein is included a mechanism for the provision of torque on the door when it is in any postion other than closed, so as to cause it to become closed automatically.

(2) A door hinge according to claim 1 wherein one portion of the hinge is located on the door frame and a second portion is located on the door itself.

(3) A door hinge according to claim 1 in which the whole of the automatic closing mechanism is located on the door.

(4) A door hinge according to claim 1 in which the whole of the automatic closing mechanism is located on the door frame.

(5) A door hinge according to claim 1 in which a first part of the automatic closing mechanism is located on the door and a second part of the said mechanism is located on the door frame.

(6) A door hinge according to any of the preceding claims in which the automatic closing mechanism incorporates a spring.

(7) A door hinge according to claim 6 wherein the said spring is made of metal.

(8) A door hinge according to claim 6 wherein the said spring is made of material other than metal.

(9) A door hinge according to claims 6, 7 and 8 wherein the said spring is of spiral construction.

(10) A door hinge according to claims 6, 7 and 8 wherein the said spring is in the form of a cantilever.

(11) A door hinge constructed and arranged substantially as herein described with reference to and as shown in any of the accompanying drawings.

The genealogical diagram is as shown in figure 9.3.

The patent examiner looks at claim 1 and sees whether the applicant's claim includes a device already known or in use. If it does, he rejects claim 1 in view of 'prior art'. The simple example here is obviously only meaningful if it had been filed about the year 1800, for the idea of a spring-held door is trivial, but even so, claim 1 could include a trap door in a floor, in which gravity alone shuts it every time it is released. Claim 1 is therefore invalid.

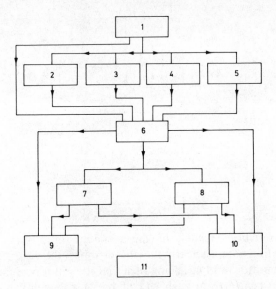

Figure 9.3   Block dependence table for the claims list in the hypothetical hinge patent

Claims 2, 3, 4 and 5 are not as broad as claim 1, but one or more of them may be rejected if the examiner can throw up any reference in an earlier patent or in any published material, or can prove that it is in common usage.

So the battle continues until the examiner reaches a claim that he allows. Subsequent dependent claims are then almost certain to be allowed, but their purpose is to add to the specification to prevent variants becoming the subject of further patents by other parties. Even in 1·800 the writer of these claims was not really hoping to get away with any of claims 1 to 5. They are an attempt to see if he can really collar the market. His real invention (in this example) is the use of a spring and he pins high hopes on being allowed claim 6, after which the remainder will stand up. But he missed out a lot, assuming claim 6 to be successful. For example he could have had a separate independent claim covering the use of a pneumatic cylinder in place of the spring, an idea that would have strengthened claim 5 perhaps to the point of acceptance.

An actual example will now be quoted together with its appropriate 'genealogical' or dependence table as shown in figure 9.4. The invention relates to the back-to-back oscillating motor described on page 148.

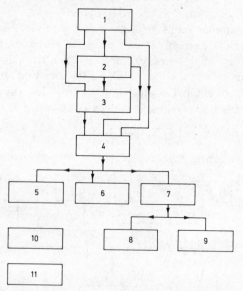

Figure 9.4   Block dependence table for an actual patent example

WHAT WE CLAIM

(1) An induction machine having a wound linear stator member co-operating with a linearly moving member, the electrical connection of the stator windings being such that the stator field moves inwardly to the centre from each end of the stator member to enable the linearly moving member to perform a reciprocatory movement of constant amplitude in which the magnetic attraction between the stator is reduced by providing a second wound linear stator member arranged parallel to the first stator member, the linearly moving member performing its reciprocatory movement in the space between the two stator members.

(2) An induction machine as claimed in claim 1 in which the linearly moving member is arranged so that it is positioned nearer to the upper stator member than the lower stator member.

(3) An induction machine as claimed in claim 1 and 2 in which the linearly moving member consists only of non-ferrous conducting material.

(4) A weaving loom employing an induction machine as claimed in any of the preceding claims for effecting shuttle propulsion in which the shuttle consists of a thin slab of non-ferrous conducting material having a portion projecting from the stator members for carrying the shuttle bobbin.

(5) A weaving loom as claimed in claim 4 in which horizontal stabilisation of the shuttle with a straight reed is effected by offsetting the thin portion of the shuttle with respect to the two stator members.

(6) A weaving loom as claimed in claim 4 in which horizontal stabilisation of the shuttle in a loom operating without a reed is effected by the provision of two separations between the windings on the lower stator member.

(7) A weaving loom as claimed in claim 4 and employing a rotary mechanism operated step-by-step in synchronism with the shuttle to form the shed and to beat the woven cloth in which means are provided which are effective when the shuttle reaches the shuttle box to convert the energy of the shuttle into energy suitable for effecting the step-by-step action of said rotary mechanism.

(8) A weaving loom as claimed in claim 7 in which each shuttle box is provided with a ratchet device which is arranged to be operated by the shuttle as it enters the box.

(9) A weaving loom as claimed in claim 7 in which each shuttle box is provided with a magnet mounted on a spindle suitably mounted for movement within the shuttle box, the arrangement being such that at the end of a pick the non-ferrous portion of the shuttle passes between the poles of the magnet and serves to move the magnet and the spindle to enable the step-by-step action of said rotary mechanism to be effected.

(10) An induction machine substantially as described with reference to figure 6 of the accompanying drawings.

(11) A weaving loom substantially as described with reference to the drawings accompanying the provisional specification and the accompanying drawings.

PATENT LANGUAGE AND STYLE

The aim of the patent writer is to obtain maximum coverage for his idea and to be as unambiguous as possible. Now commas are a hotbed for ambiguity and are therefore avoided if possible, and the style is not unlike the old 'Authorised Version' of the Bible in this respect. But unlike the Bible, the sentences can become very long, the subordinate clauses built one upon the other until the reader is inclined to exclaim in frustration (as once did a conference interpreter for a British delegation when a German delegate had addressed the company for ninety seconds without a word from the interpreter), 'For God's sake man, the verb!'

Then something has to be done to shield the writer and examiner alike from having to describe processes such as that of electromagnetic induction before he can make his new point, for this would entail patent specifications running into thousands of pages. So all agree to accept a phrase in such situations which alleviates the suffering. The magic phrase usually ends a sentence thus: 'As will be apparent to those skilled in the art.' What a mass of difficulty *that* removes! Other 'blanket' phrases to avoid lengthy description are 'means for' as in 'means for supporting', 'means for rotating', 'means for moving', and so on.

Words like 'several', 'many', 'a few' and so on lead to ambiguity and the patent lawyer will have none of it. One is 'one' and any more is a 'plurality', unless two can be described as 'a pair'. The legal side of the business leads to the very frequent use of 'the said', 'the aforesaid' and 'hereinbefore defined', to ensure there is no misunderstanding when referring back to something just described. 'A second ...' is always used in preference to 'Another ...' in which case the first reference is usually prefaced by 'A first ...'.

One of the things an inventor fears is that someone will take his patent, change one tiny detail (for example, if something is said to have to be 'at right angles', the smart boy will change it to 'between 85° and 95° excluding 90°') and be free of restriction. Although the example just chosen would rarely be upheld legally, the writer nevertheless often gets as wide cover as he can by the use of phrases such as 'substantially coaxial' (even though he knows it will always be made as exactly coaxial as is possible), 'at or toward one end' (when it will always be *at* the end), and so on.

THE FASHIONABLE FORMS OF PRESENTATION

Legal language has always appeared to lag on the language of the common man. Fashions change however, even in legal language, and often modernisation loses some of the delicacy and eloquence of that which it replaces. One need only refer to that famous passage from I Corinthians (*13, v.11*) which, in the 'Authorised Version' reads: 'When I was a child, I spake as a child, I understood as a child, I thought as a child.' All that the New English Bible can make of this is: 'When I was a child, my speech, my outlook, and my thoughts were all childish.' It contains all the facts but who on earth wants just the facts! *Ersatz* was a German word for such things, a much used word in World War II, to describe manufactured articles reduced to their basic requirements and having a dullness from which we never fully recovered.

So too, in patent literature, we have lost some of the unnecessary, but delightful style. It was customary in 1844 for example, to begin a British Patent on page 1 with words such as those of *BP. 10, 196*: 'TO ALL TO WHOM THESE PRESENTS SHALL COME, I, Joseph Meeus of Ludgate Hill, in the City of London, send greeting'.

What better phrasing could be devised to put the reader at ease and favourably disposed toward the inventor? The modern equivalent is still in old-fashioned language, but has just a touch of the New English Bible 'disease':
'We, ERIC ROBERTS LAITHWAITE of 'The Circles' Wentworth Close, Ditton Hill, Surbiton, Surrey and HUGH ROBERT BOLTON of 21e Sheffield Terrace, London W8, both British subjects, do hereby declare the invention for which we pray that a patent may be granted to us and the method by which it is to be performed to be particularly described in and by the following statement :–'

Pardon me, whilst I touch my forelock!

IDEAS AND APPLICATIONS

Two aspects of patent law that, at first sight, are very straightforward, yet lead to some of the most intricate of paradoxes may be stated as follows. In each case we give an example that makes the law at once as obviously necessary as the things we might label 'common sense'.

(1) You cannot patent an idea in the abstract.
(2) You cannot patent a new application of an old device.

An example of (1) is that you may not set out a claim that states merely: 'What I claim is an internal combustion engine that uses water as a fuel.' Otherwise a multitude of people would be filing this broad type of claim in the hope that someone will invent it within fifteen years and then be subject to their patent. Despite this, there have been many claims for perpetual motion machines which, although they were based on hardware that *we* know would never work, would doubtless have formed the subject of a court action had it not been so and someone really had achieved this subsequently.

The trouble with perpetual motion patents is that the last one to be filed in Britain was dated 1856. Unfortunately for many of us (academics) and for patent agents in particular, the applicant had sufficient private funds to pursue his originally declined applications for Letters Patent to an appeal to the House of Lords, which was upheld. We are all fallible at one time or another!

The 'application'-type claim is extremely common. A man invents the zip fastener. As soon as his work is available there is a flood of applications such as

'I claim a zip fastener for ladies dresses.'
'I claim a zip fastener for suitcases.'
'I claim a zip fastener for men's trousers.'
'I claim a zip fastener for pillow cases.'
The stream would be endless.

Now what is far *less* obvious is the extension of the idea of application of a new technique to an old piece of hardware, to the idea that a patent is no more than a combination of known articles. A man conceives the idea of using a row of electromagnets, switched on in a sequence, to drag a shuttle across a weaving loom. His patent claim can be rejected on the grounds that he has merely combined the linear motor, first patented in 1845, with an automatic weaving loom, patented even earlier. One can see the justice in this and visualise the greedy kind of person who would otherwise attempt to earn a living by doing this kind of exercise once a week. But where do we draw the line? It is surely hard on a man who, let us imagine, had invented and perfected a system of steering for a sea-going hovercraft by means of the release of jets of high pressure air from various points on the vehicle, if he is then told by the patent examiner: 'The hovercraft principle is well known. The reaction of an air-jet is well known. You have only combined pieces of the known art.'

Such a case is typical of the area where the in-fighting gets rough, and can be really expensive, argument and counter-argument flowing by letter between applicant and examiner – and *both* are humans and therefore *both* are fallible. One might have thought that the ideas expressed in this paragraph could be summarised simply by stating: 'You cannot patent just an idea, nor simply an application. You can only patent pieces of hardware.' This statement is quite true, but the conditions imposed on the hardware are many, and novelty is a matter of opinion only, in a large percentage of cases submitted, however 'expert' the authorities called in to give a judgment. It is, after all, just one facet of the much wider subject of the common law.

REJECTIONS AND FAILURES

A claim, claims or whole patent applications may be rejected for one or more of several reasons. There are listed in Section 32 of the Patents Act 1949, 'Grounds for Revocation'. The following is a layman's interpretation and summary of these.

(1) What is claimed has already been the subject of a prior claim in an earlier application.
(2) The person applying is not entitled to apply.
(3) The information claimed was obtained by the applicant by fraud.
(4) The subject matter does not disclose an invention. (In this context an invention is *defined* in the following terms: 'Any manner of new manufacture, the subject of Letters Patent and grants of privilege within Section 6 of the Monopolies Act and any new method of testing applicable to the improvement or control of manufacture and includes an alleged invention.')
(5) The idea is not novel. In this case the citations in support will be sent to the applicant who can then argue with the examiner.
(6) The idea is obvious ('contains no inventive step'). This can be argued all the way to an appeal court or to the House of Lords.
(7) The idea is not useful. This also can clearly create arguments.
(8) The device is insufficiently described or does not disclose the *best method* known to the applicant at the time (an applicant is not

allowed to 'hold back' any aspect until a later date that might give him advantage).

(9) The information was obtained on false suggestions. The last patent on perpetual motion to be granted was only given *because* the applicant could afford the money to take his appeal to the House of Lords, so this reason also is arguable.

(10) The device is contrary to law. A house-breaking device would not succeed, yet the rules here might change, indeed *do* change with changes in law. Methods of contraception were not allowed but are now the subject of granted patents.

(11) If the device is already in secret use.

(12) Although not specifically written in the act, rejection can be made on the grounds that the invention is merely a collection of known things. (Our comments on this are on page 133).

When a patent is granted there is no cause for celebration and the opening of champagne, for the system then operates as follows. The published patent comes under the eagle eye of an industrialist who, whatever else he might or should be, is a keen 'competitor' in every sense of the word. If he decides that the examiner has been too lenient, or 'blinded by science' and for any reason believes that a judge in a court of law, hearing expert testimony on the subject, would not uphold the examiner's view, he copies the equipment described, markets it and then sends the fact that he has done so to the patentee. Now the law is quite clear on this next point. No policeman or other upholder of the law will, by themselves, prosecute the copier. The onus is fairly and squarely on the shoulders of the holder of the patent, and he can elect to do one of two things, *viz*

(1) Nothing – in which case his patent is seen to be valueless, for it will thenceforth be apparent that the world can copy it, but the implication is that it is not *worth* copying, even in the eyes of the inventor.

(2) He can prosecute the copier. Here again one of two possible results can obtain:

(a) The court will uphold the validity of his patent and award damages against the copier, or

(b) The court will effectively contradict the examiner and all is lost.

It is said by the professionals that a patent is not a *good* patent until it has been challenged and upheld in a court of law. Because industrialists are shrewd men they are likely to have access to more knowledge on how likely they are to succeed in court, and one can well see the logic behind the 'court-tested' patent idea.

But occasionally there could be disaster for both challenger and patentee. As an example we quote now a specific case that was prepared for a court action and on which the advice of one of us was sought with a view to his appearance as expert witness for the prosecution. A French firm had patented and were marketing a product using a well-known physical principle. A relatively small British firm thought the idea was 'obvious', so copied the very last claim, that is the one described in connection with submitted drawings, and began marketing. The French firm did not prosecute until sufficient had *been* sold by the copiers that compensation would break them (the British firm) if they lost the case. I was asked to appear on behalf of the French firm.

After inspection it seemed to me that the whole idea was indeed obvious and that it was my duty to say so in court. But the British firm had copied bolt hole for bolt hole and washer for washer and the last claim *must* be seen as a valid patent. I advised the British solicitors acting for the French firm (it was a *British* patent that was being contested) that in my opinion both sides would lose if it went before a judge. 'How will he find?' they asked. 'He will find for your clients,' I replied, 'and it will break the British firm.' 'Then how can our clients lose?' they asked. 'Because,' I said, 'it will be obvious from my testimony and from the judge summing up that the only valid claim is the last one and any other firm need only change the position of one screw hole to escape prosecution. Your patent will not be worth the paper it is written on.' There are more ways of failing than by rejection.

Two weeks later I was told that the action had been settled out of court and that the British firm were now the agents for the French firm in Britain – a most sensible thing to do. The very act of employing an agent implied that the patent had strength (which it did not) and lessened the likelihood of the challenge being repeated.

Special care is always needed when prosecuting in another country. I remember a patent being contested in the USA that was won and lost on a single diagram. I knew the British patentee well and he advised me afterwards always to submit lots of drawings, which he said, were never ambiguous, whereas description in words might well be. It also emerged for him, as for me, that one difference in American patent law is that notebooks written and dated *in pencil* are admissible as evidence. My friend added bitterly: 'There would also appear to be an unwritten law in American courts which could be interpreted as: "All American notebooks are genuine by definition. All foreign notebooks are false".' While not of course associating ourselves with that remark in any way it will suffice to say: 'Always get an expert to advise you and even then, win or lose, it could be an expensive business.'

Patent law is like many other subjects in that one can go on learning for the whole of one's life. It was only recently that I discovered that if you can find a patent that is over fifty years old and appears to have 'died a natural death', you can refile it *in your own name* for £1. Think what a wealth of possibilities might thereby accrue from new materials of the 1970s being applied to patents of the pre-1920s. At this point we perhaps ought to stop writing and disappear into the vaults of the patents library for a year or two! Certainly we are now open to the kind of question that can be asked of a professional tipster at the races: 'If you are so sure of all the winners why don't you simply back them yourself and retire?' In writing a book on how to invent we must defend this question and our defence is simple – we would, but there are not enough hours in the day. 'The harvest is plentiful but the harvesters are few.'

The authors are extremely indebted to Mr Frank Cousins for providing a large amount of information which enabled them to give a layman's view of patent law.

REFERENCES

1. Gomme, A. A., *Patents of Invention: Origin and Growth of the Patent System in Britain*, Longmans Green & Co. (1946)

10 SOME OF OUR INVENTIONS

Part 1, M. W. Thring

In order to illustrate the processes that go on in the mind of the inventor, this chapter gives some examples of the way in which we have arrived at ideas and developed them to at least the first prototype working model stage.

MAGNETOHYDRODYNAMIC GENERATION OF ELECTRICITY FROM FUEL COMBUSTION

During World War II, I was working on coal combustion and I came across an old paper in the *Philosophical Magazine* which described the experimental observation that when fine silica particles were blown at high velocity in an air jet from a nozzle they carried a negative charge from the nozzle. I was looking for a way of generating electricity from the heat energy in the combustion gases from coal, which was more direct than going via steel tubes to steam, turbine blades, rotary motion and copper wires in a magnetic field. I compared the silica laden jet to the belt in a van der Graaf generator or the disc in a Wimshurst machine and immediately conceived what is now known as the EGD (electrogasdynamic generator); but calculations suggested that the microscopic currents would require astronomical voltages if one was to absorb the energy of combustion of a few tons of coal per hour. After thinking of this idea for some months I came up with the obvious next step that if the combustion gases were electrically conducting one could put the jet through a magnetic field and so produce a d.c. generator in which one would use a very high independently produced magnetic field. I found the figures here were much more attractive but of course combustion gases, even when laden with sodium (I found some work on the electrical conductivity in Bunsen flame in the *Ann. der Physik*) are still several orders of magnitude poorer conductors than copper so that this generator would have a high internal resistance. I left this subject in abeyance for seven years while I was setting up the Physics Department of British Iron and Steel Research Institute and working on steel furnaces. When I became Professor of Fuel Technology at Sheffield University I returned to it and discussed the possibility of what I called 'the direct generation of electricity by the electromagnetic braking of hot gases' in my inaugural lecture in 1954. This was probably the first open publication of the idea but I found, when the National Research and Development Corporation took out patents for me, that inventions dating up to fifty years before had implied the idea and that the Hungarian inventors Halesz and Karlowitz had worked on it before World War II both in Hungary and in the USA. They had already recognised the basic dilemma of the problem, that, even if you seed the gases with potassium, you still need a working gas temperature of the order of 2000°C to give a reasonable gas conductivity in the generator; if you are to have this temperature after adiabatic expansion over a large pressure ratio in a convergent–divergent nozzle then the temperature at the entry to the nozzle, that is, after combustion, must be several hundred degrees hotter still. Such high combustion temperatures can only be obtained by using air preheats of over 1000°C or burning the fuel with oxygen-enriched air. I then thought of the idea of a striated combustion system in which thin conducting layers of seeded gas from fuel combustion with pure oxygen were interleaved with thicker layers of lower temperature gas burnt with air which was therefore going through a much more economical cycle from the thermodynamic point of view. By this time British work on MHD-combustion cycles was being phased out by the nationalised Electricity Generating Board, and some experiments on this idea by a PhD student were not followed up in this country although further work was done in France. The Russians have a natural-gas-fired large pilot plant producing 25 MW(E) on an open cycle with high preheat.

It is likely that nuclear fission generated electricity will not provide the cheap energy source that was hoped ten years ago and in this case my invention of the surface controlled mole miner will receive the necessary finance since coal will become our main fuel. The coal-fired MHD generator (much studied in the USA) will then come into its own.

CONTINUOUS STEEL MAKING

The second main project I proposed in my inaugural lecture at Sheffield was a continuous furnace for melting and refining steel. Work on continuous casting of molten steel had been carried out in the Physics Department of BISRA when I was its Head and I could see the advantages of a continuous melting process (very low fuel requirement because of counterflow flame heating) and continuous refining (very clean metal with a low slag weight because of counterflow slag). This could also give a good supply of the high phosphorus-containing slag which is going to be in increasing demand as rich phosphate becomes more expensive. An experimental gas-fired counterflow melting furnace was built at Sheffield University in conjunction with United Steels and the furnace builders, G. P. Wincott (H. Southern), and we achieved over 50 per cent thermal efficiency in flame melting even on this small scale.

Schack built a full-size furnace in Germany using three rams to force the scrap in counterflow to the gases on three shelves and obtained a similar high efficiency, but had trouble in sealing the shelves. I set up a company to try to extend the idea to the study of the counterflow slag refining process, but could not raise enough money to build the large pilot plant needed for these high temperature experiments. Howard Worner, who arrived at the idea independently in Australia, managed to build such pilot plants in Australia and in Sweden but his work with steel has failed to convince the steelmakers that continuous processes can be more economical than batch processes. The Russians have a large pilot plant in operation at Nova Tula.

BOUNDARY LAYER PROPULSION

During the Second World War an inventor suggested the idea of a submarine with a rubber skin which was moved backwards on the outside and then drawn forwards through a large tube along the axis. It could thus slip through the water with no skin friction, but there were obvious problems of getting into the submarine and of the folding or contraction between the outside and the inside.

For many years I thought of ways of reducing the power consumption of skin friction on surface ships. First I thought of ring grooves around the below-waterline perimeter of the ship at right angles to the direction of motion. These would act like the cylinder of the Humphrey pump and take in water from the forward direction through vanes on the compression stroke and blow it out backwards on the explosion stroke, air being fed in from a turbocompressor for the combustion. The exhaust gases would leave with the backward propelled water, reducing its viscosity and increasing its momentum. Then I remembered the wartime submarine idea and realised it could be applied to a ship with a belt under the bows and stern and flat bottom – and two other belts on the sides if necessary. This however involved power for the shearing flow between the belt and the hull so I then thought of filling this with air, that is making an air bearing, and in this case one would be able to have the return belt movement in the air space below the ship.

We made a model of this in my design and inventions laboratory and it looked promising, provided that one could equalise the air escape on the two sides of the belt. Such a system, like the hovercraft, can move out of the water and go up a ramp on dry land but it has three advantages over the hovercraft: it has a much smaller air escape volume and hence blowing power; in water it is fundamentally stable because it has conventional hydrostatic stability whether the air is on or not; one can propel it in water by driving the belt backwards at a slightly greater speed than the forward speed of the ship. A research student studied the extent to which the grip of the belt on the water could be increased by putting transverse ribs and he showed that it did not matter what height h they were, provided they were not less than $4h$ apart.

This idea has not yet been tried on a full-size ship but it is likely to be developed first as a barge carrier which can replace locks and carry a barge up a long hillside or along a road to a factory some distance from the canal.

An alternative method of converting the forward-dragged boundary layer

fluid into a backward-pushed propulsive stream, is to have a propulsive
device at the rear of the water or air-borne vehicle. The simplest form of such
a device is a series of short aerofoil propeller blades found at an angle to a
driven belt across the stern of the vehicle. In the case of a submarine or an
airship the belt becomes a large ring around the tail of the vehicle. In the case
of a ship, of which models have been built, the stern carries a belt fully
immersed and as wide as the full amidships width, so that the stern need not
be tapered. Here the blades tips accelerate as they go round the drum at each
end so that they do not have the optimum velocity at all points. In all cases
the propeller belt can be very usefully combined with ducted jet propulsion,
which increases the thrust significantly by placing the propeller blades in the
throat of a venturi so that the forward suction on the contraction more than
offsets the backward drag on the expansion.

Another idea which follows from this line of thinking is to combine
boundary layer propulsion with the whole length of an airplane lifting wing.
Figure 10.1 shows one way of doing this with a sliding vane compressor
(with axle running along the wing length) drawing air in along a slot just at
the bottom of the front edge so that the flow along the lower surface of the
wing is reduced. The compressed air goes through a long combustion
chamber, expands in a sliding vane turbine which drives the compressor
with belts and then emerges as a high velocity jet which curves around the
back flap of the upper surface by the Coanda effect to entrain a considerably
increased flow of air over the top surface. This increases the lift and enables
a much thicker wing to be used without flow breakaway. It will also give
this circulation around the wing when the aircraft has no forward velocity

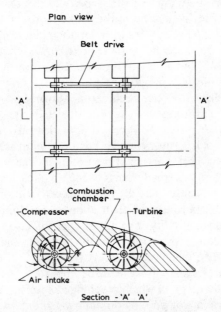

Figure 10.1   Combination of lift and thrust in a wing

141

by inducing a backward velocity over the whole upper wing surface, thus providing lift at relatively low forward velocities.

A final idea is the possibility of producing a man-propelled aircraft with a biplane having short propeller blades on a belt running along the back of the upper and lower wings. The extra lift at low forward speed, due to the combination of propulsive flow over the upper surfaces, might make a man-propelled plane just feasible.

WALKING MACHINES

About fifteen years ago, when I was toying with the idea of a domestic robot which could be trained to do routine operations in the home, I decided to tackle the problem of a machine that would climb a flight of stairs designed for human feet. I soon realised that such a machine had a much more urgent use as a powered chair to carry cripples up and down stairs.

At first I tried devices with two flat feet that could be moved forwards and upwards relative to each other by pistons and compressed air; these could be made to climb stairs but were very complex to control. Then I tried various crank mechanisms and a wheel with six projecting spokes which were withdrawn by a cam to eliminate the vertical oscillation as the weight was taken on each leg. Finally I realised that no wheel was safe on a staircase if the rim could slide on the corners of the stairs – that would be like climbing a staircase on skis.

Thus I arrived at the concept of a four-wheel-drive stairclimbing carriage with each wheel having twelve radially sprung spokes sticking out so far from the rim that the corner of the step could not reach the rim. Such a wheel puts the rubber-coated tip of the spoke on to the flat part of the step and climbs by rotating about this tip with no 'inclined plane' effect tending to cause it to slip down. If the tip of a spoke comes exactly on the corner it rolls on if going up, but rolls off and bounces on to the next spoke when going down. If the spring strength and prestressing is right there is relatively little vertical oscillation. I have been up and down stairs many times on the two battery-driven prototypes we have constructed in the Design and Inventions Laboratory (figure 10.2). We have now designed a model with three wheels; this enables it to steer round a bend in a staircase which is not a flat surface. The front wheel is driven at the arithmetic mean of the speed of the two back wheels and has small wheels at the tip of each spoke so that it can roll freely sideways. In this way, by running one back wheel backwards and one forwards, one can rotate the chair about the midpoint of the back axle to give good manoeuvrability. The new model will also have powered rotation of the seat about a low axis so that the person's weight is always nearer the upper wheel than the lower one.

The same idea of radial spokes can be applied to a tyre-less car wheel, which can be made very stiff when travelling at speed on a smooth road and very soft when braking or running on bumpy ground. In this case the wheel consists of twenty-four radial air cylinders with pistons and rods carrying

(a)

(b)

Figure 10.2 (a) *Battery-driven stairclimber;* (b) *steerable chassis for a stairclimber*

hard rubber pads to form a nearly continuous rim. A small compressor coupled to the engine, for example, a fifth cylinder on the crankshaft, provides air through the axle at a pressure sufficient to carry one-quarter the weight of the car on a single cylinder so the wheel is normally hard, but when braking the air pressure is released from all four wheels so that the wheels go flat on about six pads. The pads are spring pivoted so that a continuous flat surface is presented to the road.

143

Another type of walking machine I invented was the 'Centipede' (figure 10.3). I decided that the ideal walker for rough country and irregular obstacles would be a device which put legs vertically down in front, ran over them and then lifted them up at the back. I thought of all kinds of complex four-bar linkages to produce such a motion, but eventually realised that the motion I wanted was essentially that of a link pin on a chain going round two sprockets.

Figure 10.3   'Centipede' cross-country walking machine

A small model showed that the legs could be held vertical while the feet followed this path if one had two chains – one attached to the middle of a series of legs and the other attached to the top of each leg. The legs had to be cranked so that the upper ones going forward cleared the central pivots of the ones carrying the weight. A simpler version used T-shaped legs going round two hexagonal sprockets and with rollers at the corners of the top of the 'T' which ran on rails. By springing each leg one could have a smooth ride on the rails, even though the legs were going over quite rough ground.

I was asked to solve the problem of a vehicle to carry 1-ton tree trunks over ground and up hills when 1 m stumps were in the way. By replacing the spring legs on the centipede with soft elongated balloons, which could readily bend outwards but not in the direction of motion, it could easily move over the stumps, without sinking into soft ground and also obtained a good grip. To produce a very low power tractor with as much traction as a pair of horses I am now working on the design of a tractor with two front wheels steerable but two pairs of hind legs so that one on each side is always on the ground (figure 10.4).

In the first stage of an attempt to make a powered artificial leg I analysed the essential mechanism of human walking and produced the device shown in figure 10.5 which walks on two legs by bending the knee as the thigh begins to swing forward and straightening it as it begins to swing backward.

Figure 10.6 shows a pair of exoskeleton legs whereby a person with painful

Figure 10.4   Steerable 'walking' tractor

Figure 10.5   Walking 'legs'

arthritis can walk or stand while nearly all their weight is taken on a bicycle saddle connected by the exoskeleton to metal soles under their shoes. This required a virtual axis for the thigh rotation coinciding with the thigh joint, which I produced by having a pair of rollers on a 'T' which ran in a curved slot rail fixed under the saddle, with its axis in the required position in the body above the saddle.

An idea which I have worked out on paper and by order of magnitude calculations is to run a train on ordinary rails using air bearings in place of wheels. The air would be supplied by a small compressor to a series of holes in a plate of length 3 m and width equal to that of the rail, rigidly fixed to the bogey. When the train was stationary and no air pressure was on, this

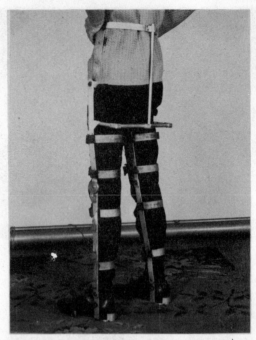

Figure 10.6   Exoskeleton 'legs' for arthritics

plate would rest on the rail, but when the air pressure was applied the plate would lift several centimetres above the rail to a height at which the air pressure between the plate and the rail just balanced, the spring strength of the steel plates forming skirts rubbing each side of the rail. At this height the air would begin to leak out and so reduce the air pressure in the space between plate and rail making a stable system.

TELECHIRICS – REMOTE MEN

My work in this field has not yet reached the stage of hardware, probably because in the field of mining the conservative approach of the men with a lifetime's experience of the old way of doing something forms an almost insurmountable barrier. However in this case necessity will, before long, have to be the 'mother of invention' and it is certain we shall be working on the mole miner in a few years because it is becoming increasingly essential as oil becomes scarcer and coal more difficult and dangerous (silicosis) to win.

Basically the idea is to produce a remote mechanic which can be so well controlled by a man in an office at the pithead that he can do assembly, fault diagnosis, repair and maintenance, and take any measurements as though he were at the point underground to which he has sent the mechanic. Once we have produced such a skill at a distance, we can develop methods of winning coal far under the sea, in very thin, very deep, steeply sloping or

146

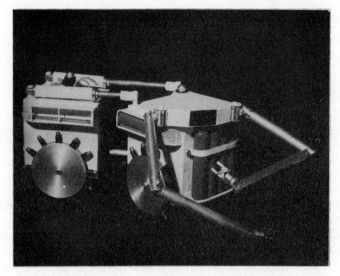

Figure 10.7   Model of telechiric miner

heavily banded seams. We can also mine uranium without men getting
lung cancer, and asbestos without men breathing asbestos dust. Figure 10.7
shows a model.

Fifteen years ago I was working on, and patented, a smokeless domestic
stove to run on bituminous washed singles. In conjunction with a Tipton
foundry we had ten stoves being tested when the coal board prevented further
experiments by saying they could not spare washed singles for the domestic
market.

Part 2, E. R. Laithwaite

There can hardly be a better illustration of the diversity of situations that lead to an idea, a diversity that makes our task in leading others along the same paths difficult, than in the selection of inventions that I outline here. There is an example of what I call 'observation and experience'. It is a sheer waste for a person to notice something which is accepted as merely 'remarkable', when had that observer spent a lifetime in the building trade, he would have seized upon it and made it the basis of a whole new family of structures. Then there is a curiosity-induced invention, a 'born-of-necessity' invention, an example of the theory coming *before* the practice (which is rare) and an example of perhaps the most difficult of all, where the observer notices the absence of something that should have been present, for this is so much more difficult than noticing something new – something extra.

My first 'professional' invention, that began to crystallise whilst I was an undergraduate, was unusual in that it was of the 'theory before practice' kind. Having had a 'double-go', so to speak, of instruction on the induction motor (first at Regent Street Polytechnic, London as part of an intensive course for other ranks, prior to commissioning in the RAF, and later as a somewhat 'long-in-the-tooth' ex-service undergraduate at Manchester), I came to regard it as something special among electric machines – a view that I have never relinquished, incidentally.

THE SELF-OSCILLATING INDUCTION MOTOR[1,2]

Living as I did both in my schooldays and in the immediate post-war years, in a cotton-weaving community, the most important ingredients in this invention were that my best school friend was a member of a family devoted to textiles. In particular his father was a man whose first aim was to stay in front in his own profession and who had at the age of fifty a thirst for new knowledge such as I have rarely found in an undergraduate. My friend was telling me that his father had just bought a loom that incorporated an entirely new concept in shuttle propulsion. Instead of the usual mechanical impact, the shuttle was blown across the loom and back by jets of air. I was shown this machine in the days when all looms were driven from belts and shafts coupled to a huge main-driving steam engine. I asked how the air jets were controlled and was told that it was by means of electrical relays. In a little more than chance remark I said: 'If you are going to bring electricity on to a loom you might as well go the whole way and propel the shuttle electrically directly.'

Even this remark had its origin further back in that I knew of the American aircraft launcher 'Electropult', and I was sure that if you could develop a 10000 lb thrust by means of a linear motor, you could certainly propel a shuttle. (This was not only my first invention, it was my first major mistake, a mistake in fundamentals, rather than in detail – the kind of mistake you try only to make once! I had neglected entirely the effect of scaling and did not know that because you can make a thing good if it is big, it does not follow that you can do the same if it is small – and vice versa.)

Having made the casual remark I got to thinking how indeed it might be done, for the shuttle has to be a freely flying member as it passes between the double row of warp threads that have been pulled apart to allow its passage (known as the 'shed' technically). The shuttle must be effectively disconnected from all objects except the single thread of weft that it leaves in its trail. Was it not traditionally known, after all, as a 'flying shuttle'? Now this last bit of information was bad for the would-be inventor. Not knowing enough about weaving I assumed the 'flying' part to be literally true, and knew that a linear motor would produce in addition to the useful thrust, about ten times as much magnetic pull, downwards, on the secondary, which was to be the shuttle, and the latter would therefore almost certainly need to have wheels which must surely damage the warp threads as it ran over them. It was years later that I came to know that the shuttle never really 'flew'. I also discovered later that some of the larger shuttles used for weaving carpets, paper-makers' felts and the like, used large shuttles that did actually run on wheels.

But I recognised a factor that I knew would put my linear motor-powered shuttle 'out of court' much sooner than would wheels. Shuttles of that day made between 60 and 150 traverses ('picks') of the loom width per minute, and the reversal of motion would have to be made by a switching of at least a part of the stator winding. Now these were the days before thyristors, or indeed any form of solid state electronics and the prospect of switching a load of several thousand volt-amps at that frequency was virtually nil. So I set about the task of investigating just *how* much of the track need be switched. The idea is illustrated in figure 10.8. When the shuttle is moving from A towards the centre, the portion PQ is such that it produces a moving field from left to right. The shuttle is therefore accelerated all the way from A to Q. What is more, the switching operation can be quite slow, because all the

Figure 10.8   *Oscillating linear motor track in which the centre section only need have its direction reversed*

time for travel to B and back to Q is available, for only on the return journey must the centre section produce thrust from right to left, to give acceleration all the way from B to P. Thus the centre section must be wide enough to make up the frictional losses per pick.

Soon after this I realised that single-phase linear motors could be just as successful as are their rotary counterparts, so the centre section could simply consist of a single-phase winding with no switching at all. The problem was solved. What is more there is even a simpler solution – extend the single-phase winding all along the track and use reversing springs at either end.

No, this was not a solution, and for two reasons, one of which is very practical and the other quite the reverse:

(1) The industry did not want it anyway (and how often have I heard this statement since that time!).
(2) The back-to-back connection of the parts AP and QB was an intellectually challenging problem, for the thrust-speed curves obtainable by induction were reasonably expressible in mathematical form and this form had just about the right degree of complexity to be analytically soluble.

In this first invention I have demonstrated how the practical need can soon recede until the elegant solution becomes an end in itself. One can easily become bogged down in a mass of analysis and waste years. Alternatively, one can, as in my case, spend the same years most profitably in obtaining a knowledge of induction motors such as I could never have obtained from books, or papers in learned society proceedings, or from courses of lectures. There is no substitute for 'getting both feet in the trough'!

The perfection of my invention is illustrated in figure 10.9. The shape of the speed-thrust curve is typical of the comparable speed-torque curve of a rotary machine and is to be found in almost every book on the induction machine. But see how it operates in my favour for the purpose I had in mind.

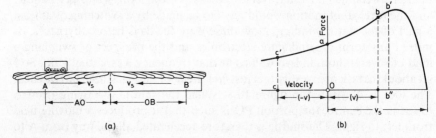

Figure 10.9   A linear motor track whose speed/force relationship is such as to give oscillation without switches; (a) track arrangement and (b) speed/force curve of the motor

Starting from rest at A (figure 10.9(a)), the secondary experiences force Oa on figure 10.9(b). As it accelerates towards the centre, the force increases as indicated in figure 10.9(b), the operating point moving steadily along the

150

characteristic from a to b. Let us assume that the secondary has reached a speed v, corresponding to point b, as it crosses the centre of the track. On entering the right-hand half, the secondary finds itself travelling to the right (due to its inertia) in a field moving to the left. So far as the coils that set up this field are concerned they have just acquired a secondary that is moving backwards (i.e.) at speed $(-v)$. This corresponds to a retarding force cc' (figure 10.9(b)) and the operating point travels from c' back to a. But notice how all the forces on this part of the curve are less than are any of the forces on ab. So it takes the right-hand half of the motor longer, in both time and distance, to stop the secondary than it did the left-hand half to start it. In other words it stops at B where OB > AO. Without a switch, the motion continues through the point a as the secondary returns towards the centre, but having, so to speak, a 'longer run at it', it attains a greater velocity b'b'', and this build-up in amplitude of oscillation will continue as long as the characteristic abb' is rising! Once over the 'hump', the accelerating forces soon begin to lose out to the decelerations, which stay more nearly constant. The diagram is more explanatory if we imagine it to be folded about the axis Oa and redrawn as in figure 10.10. The arrows now indicate a cycle of stable amplitude devoid of any switching mechanism whatever.

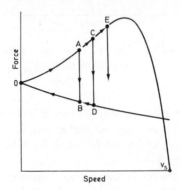

Figure 10.10—Figure 10.9(b) folded about a vertical centre-line; build-up of oscillation follows the sequence OABOCDOE, etc.

Elegant? Yes, but I couldn't get it to work! All the motors I built had speed-thrust curves of the form shown in figure 10.11, which of course yielded damped oscillations in which the secondary rapidly came to rest in the centre of the track. I looked up all the good books on induction motors and they all told me to use a large ratio of leakage reactance to resistance to get a curve shaped like that of figure 10.9. Now leakage reactance is born of magnetic circuit imperfection. R is resistance, the electric circuit's lack of perfection. It never struck me at the time that it was odd that the success of my device should hang solely on the ratio of two imperfections, the trivia of machine theory or, as the writers of the lyrics for the stage musical *Kismet* put it: 'Is fortune predicated on such tiny terms as these?'

I experimented with this back-to-back motor for over a year, at the end of which time I pushed it beyond its capabilities one day by stuffing more current through the windings than was good for them, the varnish on the wire reached flash point and the whole winding burnt out in an almost explosive manner! But I would not give up (my colleague's 'horse in a field') and wanted the motor rewound. The technician who had to do the job asked me: 'Can I use a few turns of thick wire on each coil, instead of a lot of turns of thin wire like we used before? – It was a lot of trouble winding all those coils.' 'All right,' I said, 'we can always connect all the coils of each phase in series instead of in parallel. It will make no difference.'

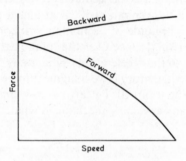

Figure 10.11   A shape of speed/force curve totally unacceptable for producing self-oscillation

Now if this had been a *rotary* motor I would have been correct in my last statement. But when it was rewound and tested, it operated magnificently. Almost any kind of secondary would oscillate. You could hardly prevent oscillations from building up!

I had had my first lesson in the *essential* difference between series and parallel connection, a difference that could be (and had been!) unnoticed by generations of machine designers in past years. And yet, the older men knew of it – the pioneers of the nineteenth century, and the Behrends, Ferraris and Heylands of the early twentieth century. To what shall I attribute this success that did so much for me at the time by way of personal 'lift', but in the later years especially in what is broadly termed 'experience'? One could say to an accident, to stupidity when I burnt out the winding, to the pure chance suggestion of an overworked technician? No there was more. There was a tenacity – a cussedness if you like – that insisted on a rewind after total failure. This invention was to pave the way for others, it was to lead to research into high speed transport, to the much more universally useful linear motors we now call transverse flux machines and to 'magnetic rivers'. My examples of the different manner in which inventions occur, the different motives, the different approaches, are almost all parts of a continuing story. Certainly this next still had the flying shuttle as objective.

152

ELECTROMAGNETIC LEVITATION

Although it was certainly the intention to tie the fruits of this invention to the textile industry, I submit it here as a product born almost solely of curiosity, for I had heard that a man had demonstrated, in a Paris exhibition of 1951 or 1952 (I never *did* trace the original source), an aluminium frying pan suspended in alternating magnetic fields and had proceeded to fry eggs in it, by the inevitable induction heating that accompanied the levitation. This I just *had* to do! But no sooner was it accomplished than I turned my thoughts to 'linearising' it, with a view to combination with a linear motor and, who knows, the flying shuttle at last?

Levitators of aluminium spheres and discs were fairly standard arrangements, such as that shown in figures 10.12 and 10.13. Two concentric coils are set into slots in a laminated steel structure (not easy to make with radial laminations). The two coils are fed with currents of different phase so that a rather poor sort of travelling field exists radially inwards, to prevent the sphere or disc from drifting from the centre as a result of its own 'shaded pole action'.[3] What I did with this model was possibly the first example of many mental exercises that I was to perform in the years to come, where I changed the shape of a known piece of technology to make it fit a known problem (either a real industrial problem or a self-generated 'ivory tower' type).

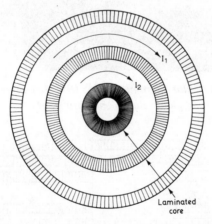

Figure 10.12   *Plan view of concentric coil system for levitation of a conducting disc*

In this case I imagined the concentric coil structure to have been put under a steam roller and 'squashed' until it took the shape shown in figure 10.14. I noticed at once with pleasure, that the coils could be divided into straight portions, in slotted, laminated iron, and bare end windings at each end, which obviously played no part in the action for the supported plate could, in theory, be many miles from either end. The laminations were easy to assemble. They were merely E-core transformer stampings.

Figure 10.13   A circular aluminium disc levitated by a changing magnetic field (this is only transient)

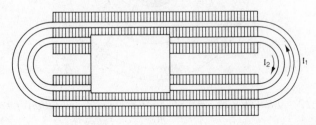

Figure 10.14   Plan view of a levitator for rectangular conducting sheets

The first machine was put through its paces, including optimising the phase angle between the currents in the two coils and we found that 180° was as good as any, so the two coils could be connected in series, back-to-back, to give instantaneous current flow directions as shown in the figure.

Almost at once it was realised that the supported aluminium plate would not be able to detect the difference between this arrangement, and that shown in figure 10.15, in which it was possible to separate the two blocks to accommodate different plate widths. This idea of crediting an inanimate object, such as a metal plate, with a personality, a cussedness, human abilities and failings was to play a much bigger part in further inventions some ten years later. It is an attitude of mind that makes the apparatus become your *adversary*, in a real sense. (But when you get to throwing things at it when it refuses to behave it is time to go home!)

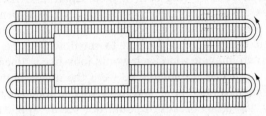

Figure 10.15   Alternative coil arrangement giving the same effect as that in figure 10.14

The rectangular plate levitator was seen to retain the same one degree of freedom as had its circular plate ancestor. The latter was free to rotate without inducing braking currents and the former could move up and down its track in like manner. Soon it was realised that it was, in one sense, *perfect*, for the plate could never be made to be at rest; any force, however small, could move it. It was a perfect measurer of levelness. In practice it was rather a measure of inhomogeneity of manufacture, but it was impressive none the less. Its simple development, as a change of shape from a known object of the same ingredients, is included here for two reasons. First it illustrates the technique without any depth of understanding being required; second, it was to play a leading part in two further inventions still to be described. In respect of the latter it should be said that it was once incorrectly connected, with one coil reversed, and this connection was found to make no noticeable difference to the performance!

'FILLING IN GAPS'

One kind of inventive process that can be formalised occurs wherever a set of items can be classified and sorted. If, for example, there is a bag of bricks of which 8 are cubes, 7 are rectangular bricks and 8 are cylinders, and there are 2 blue, 2 red, 2 yellow, and 2 green of each shape, making 6 of each colour except one, it is a fair bet that one of the rectangular blocks is missing and that the missing brick is of the same colour as that of which only 5 are to be found.

This kind of classification is useful in biology, anthropology, and other life sciences. When the first 'birdwing' butterfly was discovered it was so huge and so different from any butterfly that had been seen that it was thought to belong to a whole new order of insect. Accordingly it was given the generic name 'Ornithoptera'. Later, other species were found, and slowly it became

155

apparent that (a) they were Lepidoptera, like the rest of the butterflies and moths and (b) they were related to and ultimately proved to belong to the family of 'Papilios'. Now many of the Papilio have tails (their common name is 'Swallowtail'), and it was believed that a tailed birdwing must exist, before the first specimen of Ornithoptera paradisea (the tailed birdwing of Paradise) was captured.

Having worked on linear induction motors for some ten years at Manchester University and having seen my colleague and friend John West (now Prof. J. C. West of Sussex University) develop a form of d.c. linear motor for the positioning and control of a magnetic reading head for a computer magnetic drum store, I decided to set down all the basic types of electric motor and imagine what each would look like if 'linearised'. Among the ones that suggested no advantage were the a.c. commutator motors, both single phase and polyphase, and hysteresis motors, the former because of complexity, the latter because I could not see an application. But synchronous motors stood out as a grave omission and I set out at once to see what the topology of a synchronous linear motor might be like. Knowing that it could only be started by running it up to within a fraction of a percent of full speed by some other means, from which situation it could 'lock in', I tried a few permanent magnets let into an aluminium sheet that was the secondary of a double-sided linear induction motor, on which it could accelerate, I thought, to a sufficient lock-in speed. The first two runs wiped two sets of so-called 'permanent' magnets clean! – and they were expensive.

I decided that if continuous propulsion was not profitable, or indeed possible (please note that this was before the thyristor inverter had made its mark), I could at least try an oscillatory version. At once I saw the rotary/linear analogues as a key to the design. The maximum kinetic energy of a reciprocating mass whose amplitude is a and angular frequency ω is $(a\omega^2)$ and the motor must supply this energy and yet recover it again, twice per cycle. This giving and recovering is a power conversion, electrical – mechanical, which is always expensive in losses and makes for low efficiency. I decided that an iron magnetic circuit was too heavy to throw about and I would therefore only allow the armature conductor to move in a large enough airgap to accommodate it. Soon, with the help of John West, we had the arrangement shown in figure 10.16 which was most promising. We had isolated the alternating and direct flux magnetic circuits and yet used the same iron for the two in most of the machine. A is the oscillating loop receiving the mechanical drive (if used as alternator) or delivering it as a motor. B and C are the d.c.-carrying coils with poles opposing, D and E are the a.c.-carrying coils with poles assisting. It is a transformer with added d.c. fed poles; it was elegant and we were proud of it – we even got as far as making a 4-pole version of it.[4]

But we fell foul of a much more fundamental fact that dashed our hopes. The value of the magnetic permeability being what it is, and copper having the resistivity that it does, electric machines only work well when their linear speed is high – that means above 20–30 ft/s. To use 50 Hz in the a.c.

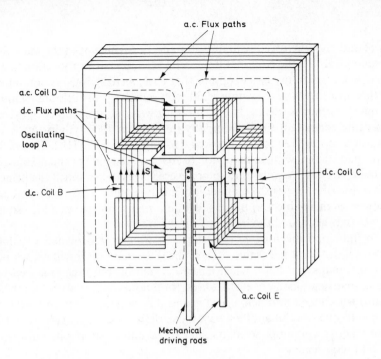

Figure 10.16 labels:
a.c. Flux paths

a.c. Coil D
d.c. Flux paths
Oscillating loop A
d.c. Coil B
S
S
d.c. Coil C
a.c. Coil E
Mechanical driving rods

Figure 10.16   Arrangement for a self-oscillating synchronous linear motor

circuit involves the use of a stroke of over 8 cm to comply with the above requirement, and even with tuning end-springs the device was a monstrosity that shook the foundations until they cracked. But it had been a useful exercise, for I discovered one of the impossible boundaries between the linear and the rotary world. A reciprocating linear motor will self-start from rest, every time, for it is not called upon to perform a stroke of maximum amplitude during the first cycle after switch-on. Fit the device with a mechanical rotary–linear converter and it is at once restricted in this respect as is the rotary machine. It is *rotation* that prevents the self-starting of a synchronous machine and not the nature of the electromagnetic mechanism itself.

As for the other gaps in the linearisation process, Professor Paul of Bangor University has perfected a most ingenious series of reluctance-type linear motors for use as actuators, whilst both linear d.c. and synchronous machines are coming back into view with the advent of solid state inverters and the present interest in linear motors and cryogenic levitation for possible high-speed transport systems.

BORN OF NECESSITY

I moved on a whole decade in linear motors. The business was getting serious. Whilst I had pursued my researches very largely on the basis of 'wasn't it all interesting?', pausing only to write up one or two of the more interesting facets in the hope that they would bring me academic promotion, I became

157

more and more involved with industry and in particular, in the mid-1960s it became clear that the National Research Development Corporation to whom I had assigned all my patents up to that date, were seriously contemplating setting up a company to levitate a 250 mph vehicle on air cushions running on a concrete beam as track. The linear motor was a natural choice of drive and I became consultant to the company involved in 1967. Almost at once an employee of the company Tracked Hovercraft Ltd (THL), Denys Bliss, and I had success with an invention to make a suspended vehicle safe even if its main electrical supply were to be severed. It is not so much the principles that I describe here as the manner in which you need a man with just a little knowledge in a subject to put his idea to a man who knows a lot – so much in fact that he missed the best idea on the way!

An induction machine can be run as a generator provided you have a means of driving it faster than its own field. In that case it will deliver power into the mains that fed it when it ran as a motor, but it cannot feed into a 'dead' load, e.g. a bank of electric fires, for it needs to 'know' (personification again) what its natural field speed is and this can only be fixed by the main supply frequency. Put another way, it can deliver *watts*, provided you feed it with reactive volt-amps to fix its field speed and to supply its out-of-phase current requirements.

Denys Bliss's idea was to run the air cushion fans from synchronous motors. Now a synchronous motor is known to be capable of delivering reactive volt-amps, provided it is supplied with mechanical drive, or with *real* watts to keep itself running. An induction machine and a synchronous machine, when coupled, therefore provide exactly a 'Jack Spratt could eat no fat, his wife could eat no lean' situation! After that it was a matter of control. Denys's patent ran the synchronous machine from an auxiliary motor whose speed was controlled from signals derived from a speed-measuring device (radar, or other similar device). He planned to switch a part of the linear motor so that it produced reversed travelling field to apply electric braking, the power coming from the induction generator portion of the linear motor, which was kept in that condition by the frequency from the synchronous machines (now running as alternators) which derived their power also from the induction generator. The feedback from the speed-measuring device ensured that the frequency was reduced in step as the vehicle speed reduced. It was a great idea, but it was complicated, and complication leads to more equipment and that in turn incurs a greater risk of failure when you least want it.

But his thoughts stimulated my own and I saw the system as a self-stabilising arrangement having no need of tacho-generator, feedback amplifier, or auxiliary drive to the fan machines. Figure 10.17(a) shows the original arrangement; figure 10.17(b) shows the system after pruning. On mains failure only switch S is operated on a 'no-volt' detector. The auxiliary machine, which *could* be the fan motor but could probably be much smaller and normally supply lighting, catering facilities, etc. has a much lower moment of inertia than the inertia of the vehicle. So the immediate reaction

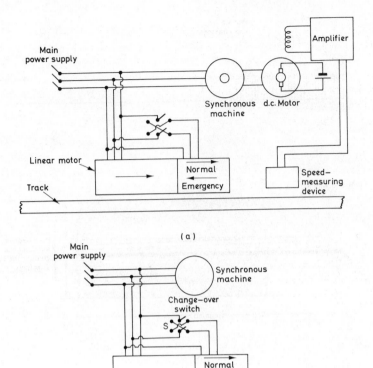

Figure 10.17 'Fail-safe' systems for linear-motor-propelled vehicles; (a) original scheme and (b) simplified arrangement

of this small machine is to slow down, bringing down with it the frequency of the reactive volt-amps that magnetise the linear motor, so the unswitched part generates power to keep the speed of the rotary motor only just below the speed needed for itself. The system is self-stabilising, for if the rotary machine overshoots with the speed drop, the linear generator responds with more power to speed it up a little. The lift fans can be paralleled with the linear motor.

The voltage is set by iron saturation, in the rotary auxiliary, just as it is in a d.c. shunt generator. No electronics, no additional equipment, a fail-safe system was the result of the ideas of one man and the experience of another. As other countries outside the UK began to experiment with linear motor drives for high speed vehicles, THL struggled to keep ahead in the technology, and the very existence of this company laid bare a fundamental problem with high speed linear motors. The problem was a double-headed monster, the two being illustrated separately by a plan view in figure 10.18(a) and the other by a side elevation (figure 10.18(b)).

The first picture is concerned with the electric circuit, where coils as in

159

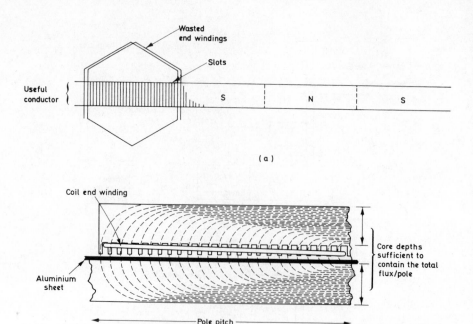

Figure 10.18   The problems of long pole pitches in linear motors; (a) excessive end winding lengths (plan view) and (b) excessive iron core depth (side view)

(a) are considered to have 'useful' portions where the coil sides are let into the slotted steel, and wasted end windings that add to leakage reactance and resistance, lowering both power factor and efficiency. What is more they protrude perhaps very far beyond the edges of the motor proper, taking up valuable space and contributing nothing. If the pole pitch is a foot and the motor width a foot, the situation is not serious. But when fed from a 50 Hz supply, the field travels the span of one N-pole and one S-pole in $\frac{1}{50}$ second, in this case, 2 ft/$\frac{1}{50}$ second or 100 ft/second ($\approx$67 mph). For 250 mph, the pole pitch rises to over 4 feet (allowing for slip) and the situation becomes intolerable. To some extent this can be overcome by short-pitching the coils, which only adds to the ohmic loss in the windings. Even reducing to $\frac{1}{3}$ the span only doubles the loss, so this part of the problem is not insurmountable. But the magnetic circuit (see figure 10.18(b)) is another story. The whole of the flux from the end pole of a linear motor of that size must pass along the steel at the back of the primary slots, and even though each tooth of the primary be only $\frac{1}{3}$ the width of a slot, a pole pitch of 4 feet demands an iron core depth of at least a foot. The track cost alone is prohibitive.

Was all my work to be wasted? Was this the rock on which we were to founder? The managing director of THL (Tom Fellows) went to investigate

160

what other countries were doing about this problem. He found that in the USA the solution being sought was to raise the frequency of supply to the linear motor by on-board power processing equipment that weighed 17 tons (for a 3-ton motor on a 50-ton vehicle!) The UK solution was much the same but the on-board inverter weighed 13 tons. It was just 'not on'. In France, he found that they had resorted to the old-fashioned, so-called Gramme-ring windings shown in figure 10.19(a) and linearised in figure 10.19(b). But it was soon found that such machines had power factors of the order of 0·05 – the 'physician' had healed the healthy circuit and left the other to die!

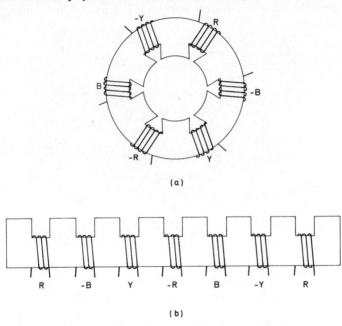

Figure 10.19 *'Gramme-ring' windings for induction motors; (a) rotary and (b) linear*

Looking back it seems incredible that we should have taken so long to find the real solution. If you change something because it is better than merely doing nothing, and observe that what you have changed has made it worse, there must be a high probability that doing the opposite will make it better. But it took me six weeks. During that time I kept repeating to myself: 'The French made Gramme-ring windings and made it worse.' It did not suggest anything to me. I grew more and more tired and red-eyed until I could never be sure whether I was awake, or asleep and dreaming. Often, as I was going to sleep proper, at night, I would have the most brilliant ideas and in the morning I was able to recall them – and always they were rubbish – the products of a dream world alone. But one night, the 'dream thought' was: 'The French turned their electric circuit through a right-angle and made it worse. Why don't we do the same to the magnetic circuit and make it better?' In the morning I thought, 'What rubbish did I think up *last* night – but wait a minute, that wasn't rubbish. That was IT!' Notice the change in wording that led to the secret. A hundred times I had said 'made Gramme-ring windings' – and that told me nothing. But I had said instead 'turned their electric circuit through a right-angle' – and that made me aware of what I now call 'three-dimensional engineering'.[5] Figure 10.20 shows the simple thing I had done. Put two linear motors side by side with N and S poles opposite each other everywhere. All the magnetic flux now goes transversely and the core depth is independent of pole pitch, the electronics are unnecessary, the track material is minimal. The age of the transverse flux motor was born.

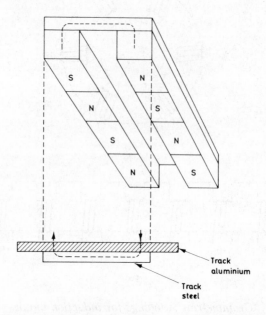

Figure 10.20   Basic arrangement of a transverse-flux motor

We are given to understand from television detective plays that the police ask witnesses to re-tell their stories time after time after time. Now we know why. They are looking for that slight alteration of a word or two that is the vital bit. One cannot be an inventor and be complacent. Transverse flux motors were a whole family of machines, one of which became an incredible extension of the idea.

THE LUXURY INVENTION

I had long known that I work best under pressure and when there is a target date to meet. Early in December 1971, an international exhibition was announced to take place at Dulles, near Washington, USA in May 1972, named 'Transpo 72'; each organisation participating in what I now called

'the High Speed Transport Game' (for Game it was, or politics, or something, but not science!) was to be allocated a space 30 ft by 20 ft within
which to demonstrate a working model of a system.

To have a 30 ft run built commercially we needed to have the prototype
built by January 1972, if we were to meet our target date. Tom Fellows
began by suggesting that the American nation as a whole were so panic-
stricken about pollution that we should go for an all-electric solution, and
pointed out that we already had the levitation part (figure 10.15) and that a
linear motor up the centre of the track was all that was really needed. But
he also said that a model vehicle longer than 3 ft would look ridiculous on a
30-foot track. A 3-foot long vehicle, to scale, was 4 in wide, 5 in with a bit
of imagination. The track of figure 10.15 was 10 in overall and could only be
run for 30 seconds before burn-out. To reduce all linear dimensions by a factor of 2 would probably cut the running time to 2 seconds – the fight was on!

It was now my turn. Looking at the cross section of the levitator (see figure
10.21(a)) and still regarding the aluminium plate as a living creature, I
argued that it probably could not 'see' the outer edges of the primary steel

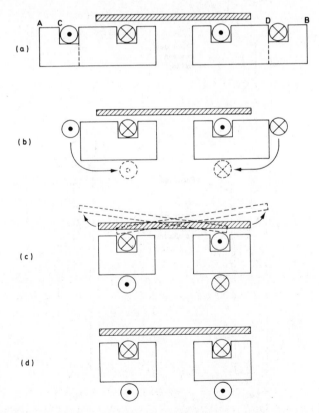

*Figure 10.21   Stages in the development of a 'magnetic river' (all figures
are lateral cross-sections)*

(A and B). 'It is all a question of edges in this game,' I argued. 'If it can't see those, it probably can't see C and D either, so we have eliminated 1 in of outer tooth ($\frac{1}{2}$ in per side) (figure 10.21(b)). But electromagnetism is only a question of linkages, so we can move the outer limbs of both coils underneath, and save another inch and a bit (figure 10.21(c)).' We tried it, and the plate always tipped to the position shown dotted. It was totally unstable. Then Fred Eastham (now Professor J. F. Eastham, Aberdeen University) changed over the connections, and the 'magnetic river' that was capable of lifting, guiding, and propelling, all from one and the same set of coils, was here to stay (figure 10.21(d)). We did even better before Transpo 72 by reducing the system to just one of the two blocks shown and still retaining stability, and I reckon we 'won' Transpo 72.

But the storm clouds were gathering for THL, and closure came, followed by the giving of evidence before the Select Committee of the House of Commons, who soon asked to come and see our work at Imperial College. There was a new target date, a new challenge. We knew that our magnetic

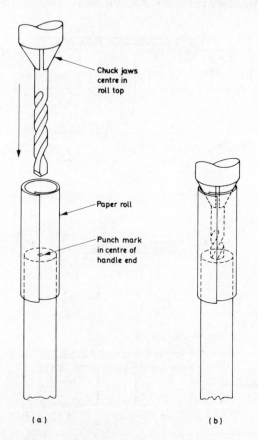

Figure 10.22   Method of drilling an axial hole; (a) before insertion of drill and (b) drilling in progress

river was still a museum piece, for 95 per cent of the input power went into lift and guidance, and a motor with a maximum efficiency of 5 per cent is nearly useless. Before the Select Committee arrived Fred Eastham had designed a winding that demonstrated clearly the possibility of a magnetic river with over 95 per cent of the input going into mechanical driving power.[6]

The man who replaced Fred Eastham following his appointment at Aberdeen was Dr Ernest Freeman and within the year he and a colleague[7] had extended the magnetic river technique to a point where lift and guidance could be made independent of speed (unlike thrust which varies with speed). Fred Eastham had invented the tubular TFM. Three-dimensional engineering is now in fashion so far as electromagnetism is concerned.

For a complete change of subject, from the complex to the simple, in fact easily the simplest of all my inventions, I need no more than a sheet of newspaper. This therefore is an invention that could not possibly be patented, alas! Once upon a time I hoped it might win a magazine competition for 'do-it-yourself' ideas, but the opportunity never presented itself, so here it is, free of charge, for the amateur carpenter. The problem is to drill a hole axially down a broom handle with great accuracy, using only an ordinary brace or wheel brace or at best a hand-held electric drill.

First mark the centre of the end of the handle, for here you can place the point of the electric drill. Next wrap a sheet of paper around and around the handle, making many layers, as shown in figure 10.22(a). Secure the end of the roll with sticky tape. Now slide up the paper tube to the position shown in figure 10.22(b) so that the upper end contains the drill chuck as shown. A cardboard tube (for this is what you have effectively made) is very resistant to bending (so is almost any kind of tube for that matter) and as you drill, the chuck will be laterally centred, and the bit, started at the centre, will be kept axial. The tube will sink naturally as the hole gets deeper and will be as well centred as if the handyman had had the use of a lathe!

REFERENCES

1. Laithwaite, E. R. and Lawrenson, P. J., 'A Self-oscillating Induction Motor for Shuttle Propulsion', *Proc. I.E.E.*, **104A**, No. 14, 93–101, April (1957)
2. Laithwaite, E. R. and Nix, G. F., 'Further Developments of the Self-oscillating Induction Motor,' *Proc. I.E.E.*, **107B**, No. 35, 476–486, Sept. (1960)
3. Laithwaite, E. R., 'Electromagnetic Levitation', *Proc. I.E.E.*, **112**, No. 12, 2361–2375, Dec. (1965)
4. Laithwaite, E. R. and Mamak, R. S., 'An Oscillating Synchronous Linear Machine', *Proc. I.E.E.*, **109A**, No. 47, 415–426, Oct. (1962)
5. Laithwaite, E. R., 'Three-dimensional Engineering', I.E.E. Conf. on Linear Electric Machines, London, Conf. Pub. No. 120, 1–8, Oct. 21–23 (1974)

6. Eastham, J. F. and Laithwaite, E. R., 'Linear Induction Motors as Electromagnetic Rivers', *Proc. I.E.E.*, **121**, No. 10, 1099–1108, Oct. (1974)
7. Freeman, E. M. and Lowther, D. A., 'Normal Force in Single-sided Linear Induction Motors', *Proc. I.E.E.*, **120**, No. 12, 1499–1506, Dec. (1973).

EPILOGUE

TO THE READER

We hope that having read this book, you now agree with what we said in the Preface, namely that

(1) You can become an inventor and can help others to develop their inventive ability.
(2) There are very many important human problems requiring inventive solutions.
(3) A study of the methods and principles of invention can be of positive help in developing creativeness and originality in oneself.

It will give us great satisfaction if, by writing this book and telling you of our own studies and methods, we become godparents or, even better, midwives to your inventions.

APPENDIX 3-D CROSSWORD

A cubical box divided into $2 \times 2 \times 2$ compartments with a letter in each (or a forbidden 'black' compartment) can be studied more easily if it be divided into 'layers'. It is then possible to forget the thickness and draw the layers simply as in figure A.1. Clues would then be given for 'Layer 1 across', 'Layer 1 down', 'Layer 2 across', 'Layer 2 down', 'Inwards top layer', 'Inwards bottom layer'.

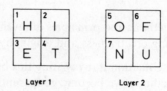

Layer 1   Layer 2

Figure A.1   A three-dimensional crossword

It is easiest to look at a solution first.

CLUES

Layer 1 across	1. American form of greeting
	3. '– tu Brute!'
Layer 1 down	1. Tarzan sometimes described as this type of man
	2. 'The thing' (science fiction)
Layer 2 across	5. 'Now is the winter – our discontent'
—	7. Phonetic unused object
Layer 2 down	5. Take a letter from one
	6. A small number, phonetically speaking
'In' top layer	1. Westward –!
	2. Kipling's famous poem
'In' bottom layer	3. A prefix
	4. The Roman 'thou'

A more complex example with blanks is shown in figure A.2.

168

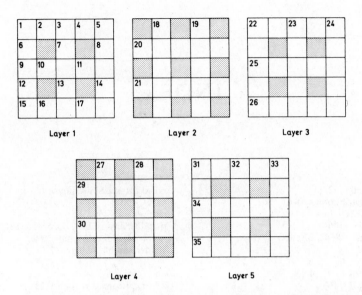

Figure A.2   A more complex three-dimensional puzzle

INDEX